PHYSICS OF THE EARTH

PHYSICAL SCIENCE AND TECHNOLOGY SERIES

Editors

Karl K. Turekian Yale University
J. Rimas Vaišnys Yale University

Avizienis	INFORMATION AND COMPUTERS
Bernstein	LAWS OF NATURE
Cameron	STRUCTURE OF THE UNIVERSE
Davidovits	COMMUNICATION
Davidovits	TRANSPORTATION
Fenn	TRANSFORMATION OF ENERGY
Gordon	PHYSICS OF THE EARTH
Turekian	CHEMISTRY OF THE EARTH
Vaišnys	CHEMISTRY AND PHYSICS OF LIFE
Wheeler	STRUCTURE OF MATTER
Willis	RELATIVITY AND UNCERTAINTY

PHYSICS OF THE EARTH

ROBERT B. GORDON *Yale University*

HOLT, RINEHART AND WINSTON, INC.

New York Chicago San Francisco Atlanta
Dallas Montreal Toronto London Sydney

ISBN: 0-03-081195-3
Library of Congress Catalog Card Number: 72-184298
Printed in the United States of America
2345 090 987654321

Editors' Foreword

Physical science and technology in themselves bring neither salvation nor doom for mankind. Rather, they are as much a part of man's cultural environment as poetry and music. The comprehension of the physical world and its harnessing are no less acts of the human spirit than the creation or appreciation of works of art.

This series presents fundamental knowledge of our physical world in as elementary a way as the editors and each author thought possible, without reducing the presentation to banality. Not all parts of the books can be read with equal ease. However, just as there is no effortless way to learn to play the violin well, there is no way to comprehend nature without applying oneself to the task. It is hoped that through these books the reader will not only discover the beauty of the physical universe, but will also learn about the processes that lead to an understanding of the universe.

The series format permits teachers and students to choose the material to suit their own aims and interests.

KARL K. TUREKIAN
J. RIMAS VAIŠNYS
Editors

Preface

Physics of the Earth deals with the physical processes that shape the environment in which we live. It aims to explain how phenomena in or on the earth—storms, earthquakes, the formation of mountains—can be understood in terms of physical principles and simple analysis. The book is intended for college students taking an introductory science course; it assumes that some instruction in basic physics precedes or accompanies its use.

The plan of the book takes account of the conflict which arises between the need to be reasonably comprehensive in scope and yet sufficiently detailed that the treatment is not superficial or merely descriptive. A rather wide range of topics is covered in the belief that it is important for the student to become familiar with the physical environment as a whole and to see how its various constituents interact. (We recognize today that much mischief in the environment has arisen in consequence of technological decisions being taken in too narrow a context, no matter how expertly made within that context.) Of the topics included a few are selected for more detailed study, not because they are more important than others, but because they are more suitable to serve as examples of how physical analysis of earth phenomena is done. For these topics it is possible to show more of the development of the basic laws and governing equations, while for the other subjects the coverage, to fit the compass of the book, must be more descriptive. The student should acquire both an appreciation for the earth as an environment, and for the methods of physical science used to describe it.

The selection of topics for the more detailed study is, of necessity, somewhat arbitrary. The choice of interface phenomena—waves, beaches, coasts and rivers—and of ocean tides is made, first, because much that is basic to these phenomena can be described with the aid of very simple calculations and, second, because most of the processes described can be easily observed by the student wherever land meets water, be it seashore, lake or river. In the solid earth, the deep oceans and the atmosphere the most interesting physical phenomena are not so conveniently exposed to view.

A book about the science of the earth must contend with a difficulty, the complexity of the observed phenomena, which can be more easily circumvented in the study of pure physics or chemistry. The working of machinery or electronic apparatus can be understood by first studying and understanding the component parts (which may be quite simple) and then examining their step-by-step assembly to achieve the desired end. Such simplification is not always possible in dealing with physical processes occurring "in nature"; it is the complex interactions between whatever simple elements that are present which is usually the source of the phenomenon of interest. The development of cyclonic storms (discussed in Chapter 1) is a good example of this. To study the physics of the earth we must approach the world as it actually is, to attempt to understand in the face of complexity. To aid the student attain first a general understanding of each of the topics discussed, a qualitative discussion is given first; detailed calculations and derivations have been placed in appendices to the individual chapters. These can be omitted by students so inclined. General appendices dealing with definitions of physical quantities, units of measure, and with the geologic time scale have also been included.

The sequence of topics is in rough order of increasing complexity. The atmosphere is considered first. It is composed of gases whose properties are simple and familiar; it is transparent and can be studied throughout its depth. The oceans, again composed of a familiar material but much more difficult to observe, are dealt with next. The tides of the ocean are singled out for more detailed study because the physical theory is complete and relatively simple. There is also opportunity in discussing tides to develop some basic relations in hydraulics useful in the following chapters. These chapters deal with events occurring at the air, water, land interfaces: waves at sea, erosion, sedimentation and the physics of rivers. The remainder of the book is concerned with the solid earth, with sea floor spreading, earthquakes, continental drift and the formation of mountains. These are topics where the observational data remain quite incomplete (even the compositions of the materials involved can only be guessed at in some cases) and where qualitative understanding, still in the formative stage, is being rapidly advanced by current research.

R. B. G.
New Haven, Conn.
October 1972

Contents

PHYSICS OF THE EARTH

1 The Atmosphere

The blanket of gases which surrounds the globe, the atmosphere, has a mass only one millionth of that of the whole earth. But life would be impossible without it and it is through the atmosphere that the most obvious day-to-day changes in our environment occur. Compared to the liquid and solid parts of the earth, the atmosphere has been subjected to detailed scrutiny: It has been examined visually from both above and below; it has been probed throughout its depth by aircraft, balloons, and sounding rockets. Casual observation of winds, clouds, and storms is confirmed by more detailed study in showing that the atmosphere as a whole is always in motion; the most characteristic thing about weather is change. The motion of the atmosphere is driven by the heat continually received on earth from the sun. Through the dynamic processes which occur in the atmosphere this thermal energy causes rain, storm, wind, fog and drought. Chapter 1 is concerned with these dynamic processes.

SOLAR HEAT

Heat, or thermal energy, can be transferred by *conduction, convection* or *radiation.* In conduction heat is carried through material, as when a silver spoon is left in a cup of hot tea. In convection hot material is transferred from one place to another: Air heated by a radiator flows by convection throughout a room carrying heat with it. But there is no matter between the sun and the earth through which the sun's thermal energy can be conveyed by conduction or convection. The earth can be heated only by radiation. Radiant transfer of heat can occur in the absence of matter because the energy is carried by *electromagnetic waves*. This is the same type of radiation as that which transmits light and radio signals. Electromagnetic waves have a characteristic *wavelength* which, for light waves, determines the color. (The concept of wavelength is developed further in Chap. 4.) Red light has a long wavelength; shortening the wavelength changes the color through red, yellow, green and blue to violet. A hot body is a source of electromagnetic waves. The hotter an object is, the shorter is the wavelength of the waves emitted by it. When the wavelength becomes short enough, the emitted radiation becomes visible and the body is "red hot." With continued heating it becomes white hot. Radiation can be felt from an object which is at less than red heat. Its radiation has a wavelength too long for it to be detected by the eye; it is called *infrared*. The familiar domestic "heat lamp" is a source of infrared rays. The sun emits infrared and visible radiation and some rays of even shorter wavelength called *ultraviolet*. The intensity of all the solar radiation received at the outer limit of the earth's atmosphere is, in thermal units, 2.0 cal/(cm^2 min). This intensity remains constant; changes in weather are not due to changes in the intensity of solar radiation.

Solar energy is not uniformly received over the surface of the earth. The full intensity is incident on the ground only when the sun is directly overhead. When it is not overhead, the intensity is reduced because the earth's surface is not perpendicular to the solar rays and because a greater thickness of atmosphere must be transversed. These factors are illustrated in Figure 1.1. In New York City, for example, 95 percent of the solar radiation intensity is received in summertime (when the sun approaches within 18° of being directly overhead) but only 44 percent in the winter, when the sun does not reach so high an altitude in the sky. (The seasonal variation is primarily a consequence of the inclination of the axis of rotation of the earth to the plane of the earth's orbit about the sun; see Chap. 3.)

Not all of the solar energy incident on the top of the atmosphere reaches the ground; some is dispersed by scattering and reflection in the atmosphere. Scattering of light by fine particles is easily observed by shining a flashlight beam through tobacco smoke in a darkened room. The smoke particles are somewhat smaller than the wavelength of the light. Very fine particles cannot reflect or refract (bend) incident light as large particles do. Instead, some of the light incident on the fine particles is re-radiated in all directions by the particles. This re-radiated or *scattered* light makes the flashlight beam appear white. Small particles scatter only blue light (shortest wavelength), larger particles scatter red light as well. That is why smoke from the front end of a cigar appears blue while that exhaled by the smoker is white: Smoke particles consolidate with moisture in the mouth to make larger particles. Air molecules scatter most effectively the short wavelength component, the blue light, of solar radiation. The sky appears blue because repeated scattering directs some of the scattered blue light down to the ground (Fig. 1.2). If it were not for scattered light the sky would appear black except where pierced by the disc of the sun. This is how the sky looks on the moon, where there is no atmosphere. When viewed through a layer

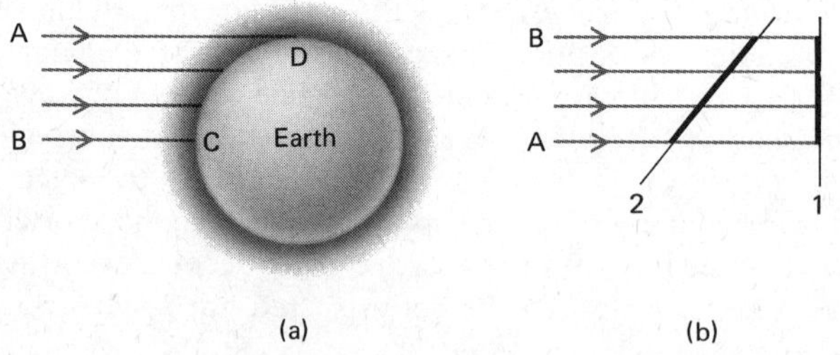

Figure 1.1 (a) Radiant energy incident on the earth's surface at an inclined angle, as at point D, is less intense because it must traverse a greater thickness of the atmosphere than that coming straight down, as at C. (b) The energy carried by the rays A–B is spread over a greater area on the inclined surface (2) than on the perpendicular surface (1) and so produces less heating per unit area.

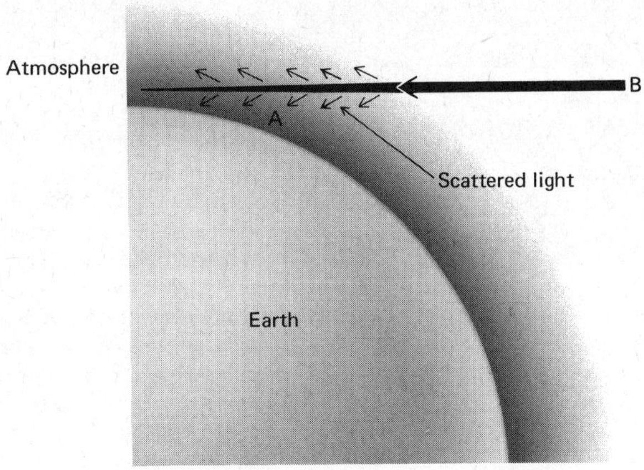

Figure 1.2 If the atmosphere were perfectly transparent an observer at A would see no light from the ray B. But blue light is scattered out of ray B in all directions. Scattered light reaching A makes the sky look blue.

of smoke and dust, the sun and neighboring sky appear red because on its path through the atmosphere to the observer, most of the blue component of the incoming light has been lost by scattering.

Water droplets and ice crystals in the air reflect solar radiation because they are too large to scatter it. All wavelengths are equally reflected. (A rainbow results from refraction and reflection of light within raindrops; it is seen only at certain angles relative to the sun's rays.) Light reflected from distant clouds appears white and sunlight received through cloud cover is simply reduced in its intensity—it appears gray. A thick cloud cover may reflect as much as 75 percent of the incident sunlight.

While particles scatter and reflect light, little of the visible and infrared solar radiation is absorbed in the atmosphere: However, ozone high in the atmosphere absorbs the ultraviolet (short wavelength) component of sunlight which is lethal to living cells.

Of the solar radiation that penetrates the atmosphere, the amount reflected at the earth's surface depends on ground conditions. Freshly fallen snow reflects 80 percent of the incident radiation; in technical language, it has an *albedo* of 80 percent. At the other extreme, a forest reflects only about 5 percent. The radiation not reflected is absorbed and if this energy were not removed in some way, the earth's surface would be continually warming up. Since its average surface temperatures are nearly constant, the earth must re-radiate much of the solar energy incident on the ground. It does this and more besides: Solar energy incident on the earth's surface averages 0.25 cal/(cm^2 min) world-wide; the energy radiated from the earth's surface is about 0.55 cal/(cm^2 min).

Despite this difference, no surface cooling is observed. The *greenhouse effect* (Fig. 1.3) makes it possible to have a higher surface temperature than thermal re-radiation would otherwise allow. The lower the temperature of a body, the longer the wavelength of the radiation emitted. The earth's surface is at a relatively low temperature and so emits much longer wavelength radiation than it receives. Ordinary window glass has the property of transmitting incident solar radiation but of reflecting the longer wavelength radiation emitted from the earth's surface. This effect is utilized in the construction of greenhouses. Within the glass enclosure radiant energy is trapped so that when the sun shines, a relatively high interior temperature results. Similarly, in the atmosphere, carbon dioxide and water vapor reflect the long-

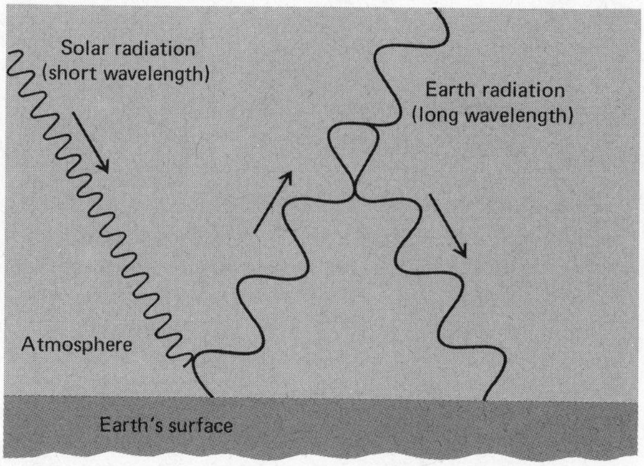

Figure 1.3 The "greenhouse effect" arises because the short wavelength radiation received from the sun is transmitted by the atmosphere whereas the long wavelength radiation emitted by the earth is partially reflected. Radiant energy is trapped by the atmosphere and the surface temperature of the earth is higher than it would otherwise be.

wavelength radiation emitted by the earth's surface. These gases act as a thermal blanket, trapping radiant energy and keeping the surface temperature well above the −40°C which would otherwise be the earth's equilibrium surface temperature. Cloud cover also inhibits the escape of the earth's surface energy to space. Clouds reflect radiation from the earth almost completely. This is why on a windless winter night, air temperatures will drop much lower if the air is clear than if there is cloud cover. Cloudy skies tend to keep surface temperatures nearly uniform by reflecting incident sunlight during the day and retaining surface heat at night.

The equilibrium surface temperature of the earth is maintained by a delicate balance among all of these factors. What is remarkable is that this equilibrium temperature happens to be in the range where reactions between organic molecules can proceed at reasonable rates; a small change in any of a number of factors would take the surface temperature out of the range where life as we know it is possible. Changing the carbon dioxide content of the atmosphere as a result of burning fossil fuels is one possible source of such change.

LARGE-SCALE CIRCULATION OF THE ATMOSPHERE

The incidence of solar radiation on the earth's surface is variable, being high near the equator and low at the poles. The atmosphere is thus unevenly heated. Since hot air tends to rise and flow over cold air, unequal heating of the atmosphere will cause circulation, just as the air in a room is caused to circulate by a hot radiator in one corner. The general circulation of the atmosphere redistributes heat over the earth's surface so as to keep temperatures more nearly uniform, avoiding climatic extremes. If the earth's surface had a uniform albedo and if the earth did not rotate but were uniformly heated about the equator, the simple circulation pattern shown in Figure 1.4 would develop. Hot air would rise over the equatorial zone and flow towards the poles at high altitude. As it lost heat along its poleward journey it would gradually sink to lower altitudes. Along the surface there would be a return flow of cooler air.

In fact, circulation in the atmosphere is much influenced by the rotation of the earth. Winds are caused by motion of air from an area of high pressure, where air molecules are squeezed more tightly together, to one of low pressure, where there is less squeezing. (In the same way air under high pressure escapes through a leak in a

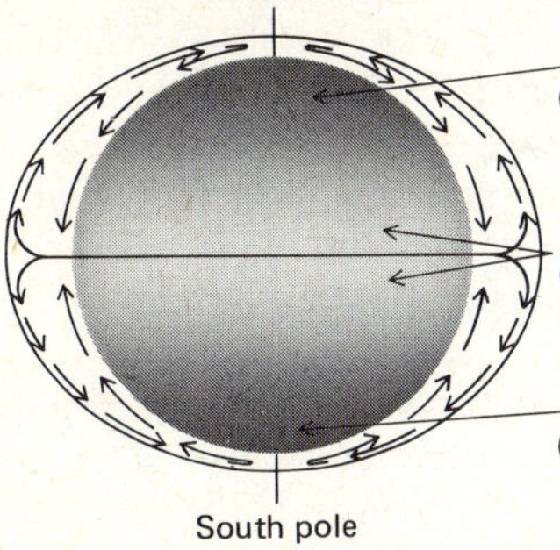

Figure 1.4 Atmospheric circulation which would result from uniform heating at the equator of a non-rotating earth. After Bowditch, *American Practical Navigator,* through courtesy U.S. Naval Hydrographic Office, 1958.

tire; flow is from the place where pressure is high to the place where it is low.) But on earth wind does not blow directly from a region of high to a region of low pressure; it is deflected because of the rotation of the earth. Suppose that the pressure is higher at A than at B in Figure 1.5a. Because of the rotation of the earth on its axis, both A and B are moving from left to right. The eastward speed of A is about 1200 km/hr; the eastward speed of B, because it is closer to the axis of rotation, is somewhat less. (If B were at the north pole its eastward speed would be zero.) Suppose that a parcel of air starts moving north because of the higher pressure at A. In addition to its northward velocity this parcel also has an eastward velocity of 1200 km/hr. It retains this eastward velocity as it moves north (the drag of ground friction is small). Therefore, as it moves further north, its eastward velocity is always becoming greater than the eastward velocity of the ground below. An observer on the ground sees this parcel of air move off to the east as a result of its excess eastward speed. To an observer on earth it is as if there were a force pushing the air to the east.

Now suppose that the pressure is higher at A than at C. A parcel of air at A is accelerated eastward by the higher pressure at A. As its eastward speed increases it acquires a greater velocity of rotation relative to the earth's axis and, hence, a greater outward, centrifugal force. The centrifugal force is shown by the black arrow in Figure 1.5b; its vertical and horizontal components by the gray arrows. The vertical component slightly changes the gravitational attraction on the air parcel but the horizontal component causes the parcel to move to the south. The force acting to the south in this case is of the same magnitude as the eastward force in the first example. This force, always acting perpendicular to the particle velocity, is called the *Coriolis*

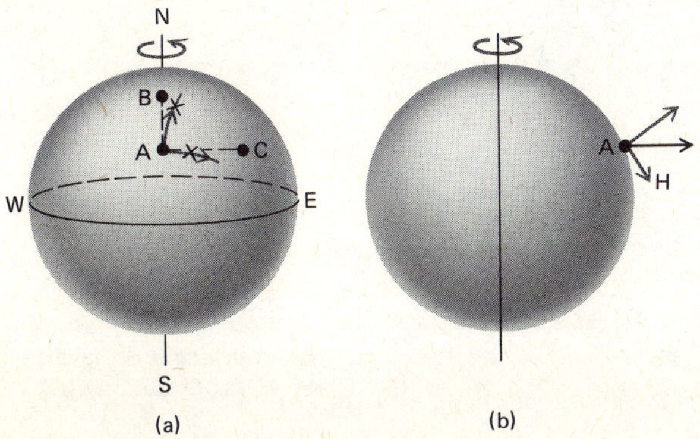

Figure 1.5 Origin of the Coriolis force. (a) When the pressure is higher at A than B air is deflected to the east because of the earth's rotation. When the pressure is higher at A than at C, the deflection is to the south. (b) The southwards acting force is the horizontal component H of the centrifugal force acting at A. (Resolution of a force into components is reviewed in Appendix I.)

force (after the French physicist who first calculated it). It acts on any object which is projected over the surface of the earth. Thus a correction for Coriolis force is required in training a cannon to fire a shell over a long distance. The result is that, in the northern hemisphere, a mass of air in motion is always deflected to the right. In the southern hemisphere the deflection is always to the left. These observations are summarized in *Buys-Ballot's law:* Stand with your back to the wind in the northern hemisphere and the low pressure is always to your left, the high pressure to your right. Check this for an observer at the two places marked x in Figure 1.5a.

The Coriolis force has a great effect on the simple circulation pattern shown in Figure 1.4 for a nonrotating earth. The north–south air flow of Figure 1.4 is deflected sideways. Consider, for example, the cool, surface air moving into the hot equatorial zone from the north. It is deflected to the right and so blows not from the north but from the northeast. This part of the circulation is known as the northeast trade wind; it is shown in Figure 1.6 which is a simplified diagram of the general circulation of the atmosphere. The trades are steady winds, long known to the navigators of sailing ships; they are a result of the strong heating of surface air in the equatorial region. Near the equator heated air is rising and there is little wind at the surface. Consequently, the zone centered about the equator is called the *doldrums*. It has always been an important obstacle to sailing ships making north–south voyages. Air entering the doldrums from the south is deflected to the left by the Coriolis force causing the southeast trade winds. These are also illustrated in Figure 1.6. It is more complex than the simple pattern of Figure 1.4 because of the Coriolis force, and also because of the seasonal variations in radiation received in the northern and southern hemispheres and the influence of continental land masses on air flow and atmospheric heating. The general circulation is governed by the equatorial low pressure (where the atmosphere is strongly heated), high pressure over the poles (where heating is weak), and zones of high pressure at mid-latitude. The circulation falls in a series of zones:

1. *Doldrums.* Upward motion of warm air over the equator results in little wind. Hot, sultry weather prevails with much cloudiness due to the high moisture content of the upward-moving air.

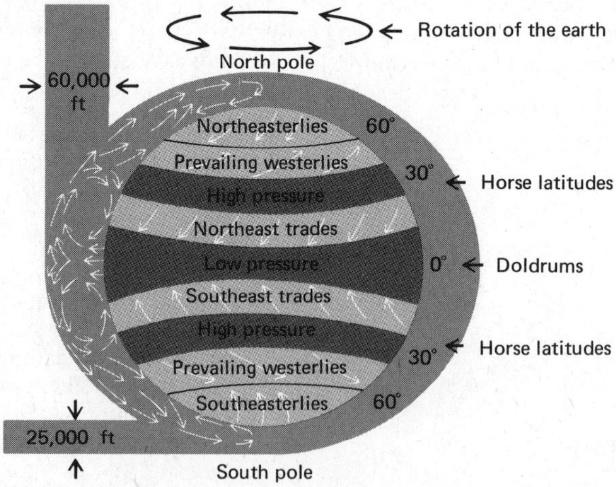

Figure 1.6 General circulation pattern of the earth's atmosphere. After Bowditch, *American Practical Navigator,* through courtesy U.S. Naval Hydrographic Office, 1958.

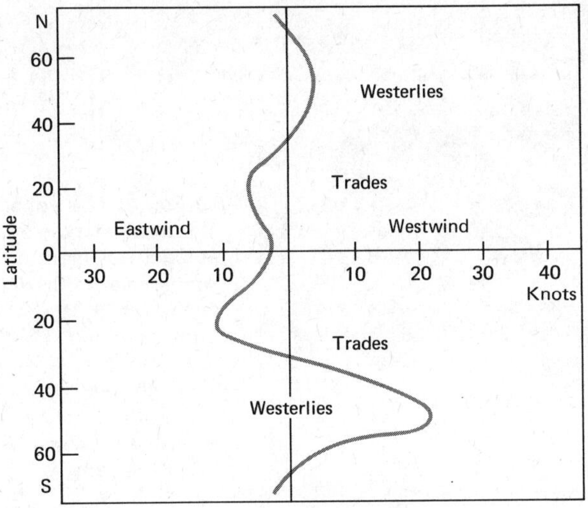

Figure 1.7 Variation of average wind velocity (in knots) with distance north or south from the equator. From *Introduction to the Atmosphere* by Riehl. Copyright 1965 by McGraw-Hill. Used with permission of McGraw-Hill Book Company.

2. *Trade Winds.* The northeast and southeast trades are relatively gentle and are among the most constant winds found in the world. The name is derived from a now obsolete meaning of trade as "course" or "direction" referring to the steadiness of their direction.
3. *Horse Latitudes.* Because of the descending cool air, surface winds are light and the weather generally fair. It is said that calm weather kills horses on a sailing ship, hence the name.
4. *Prevailing Westerlies* result from the flow of air from the horse latitudes towards the poles. In the southern hemisphere the prevailing westerlies occur over the Southern Ocean, a band of sea encircling the globe with scarcely any interruption by land. Without obstruction by land, high wind velocities develop and so this zone, centered on 40°S latitude, is known as the *roaring forties.*
5. The *Polar Easterlies* result from the outward flow of cold air from the polar regions. In the Arctic these easterlies are not strong but in the south cold air flows down off the Antarctic plateau with velocities which at times reach 200 knots.

The distribution of surface wind velocities in the northern and southern hemispheres is summarized in Figure 1.7.

STORMS

The prevailing westerlies are of special interest in the study of weather because they are the source of most storm activity. The westerlies are due to the flow of air outward from the fairly steady high pressure at mid-latitude towards lower pressure near the poles. The flow pattern of the westerlies is not one of uniform direction but of large-scale undulation around the world. One way of showing this is by tracking a volume of air which has been marked or labeled in some way. An opportu-

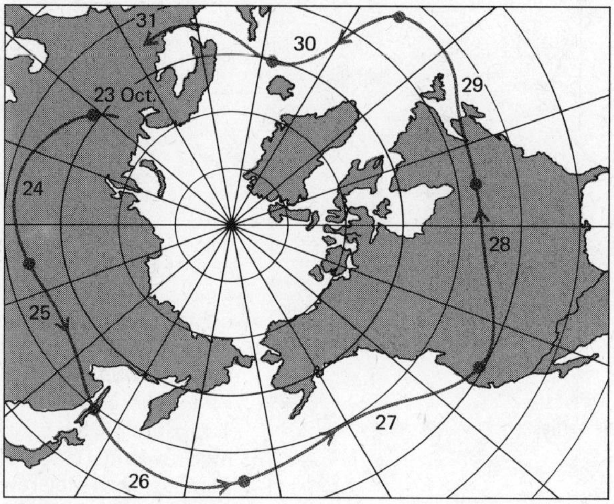

Figure 1.8 Track of high altitude air marked by radioactivity after an atomic explosion over Novaya Zemlya on 23 October 1961. From *Introduction to the Atmosphere* by Riehl. Copyright 1965 by McGraw-Hill. Used with permission of McGraw-Hill Book Company.

nity to do this has been afforded by past atmospheric tests of atomic bombs: The radioactivity released into the air at the test place can be followed as it is carried from place to place by the winds. Until the radioactivity is dispersed, it serves to mark the volume of air into which it was released. Figure 1.8 is based on data taken at the time of the Russian bomb test in the autumn of 1961. It shows the location on successive

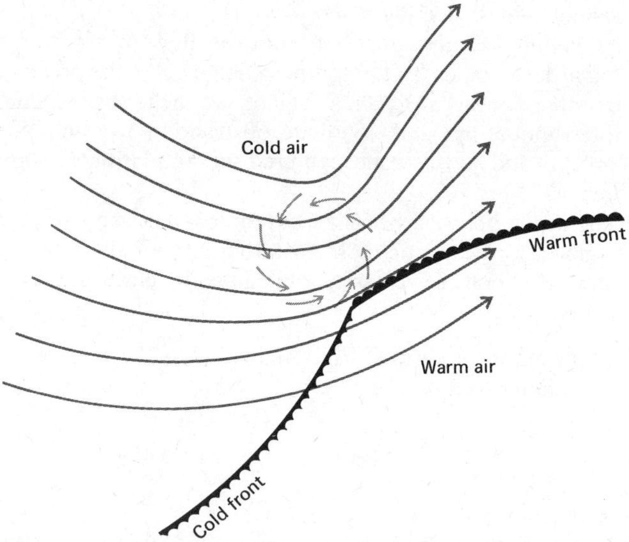

Figure 1.9 Initiation of a storm center. The path of air in the jet stream is shown by the gray lines; a developing eddy by the arrows. Normally there is warm air to the south, cold air to the north. The boundary between the cold and warm air is shown in black. When this boundary is set in motion, weather fronts develop. When the jet stream swings to the north it causes advance of warm air northwards. This is the first stage in the formation of a cyclone.

days of the radioactivity released over Novaya Zemlya on October 23, 1961. The flow of air is from west to east. It is not steady but undulatory, swinging alternately to the north and south. The high altitude component of this undulatory flow in the prevailing westerlies, not being retarded by ground friction, moves at high velocity and is called the *jet stream*. The wave pattern of the jet stream is not perfectly stable; shifts in jet stream position are the primary cause of weather changes at the surface.

A shift of the jet stream track can result in the formation of a *cyclone* or storm center. Figure 1.9 shows the situation which may develop when the jet stream swings sharply northwards. This swing brings warm air much further north than it would otherwise be and causes the jet stream to spin off an eddy, a circulating mass of air within which the air flow is inwards, towards the center of rotation. Surface air drawn into the center of such an eddy is lifted to higher altitude where the condensation of its moisture forms clouds and rain. It becomes a zone of bad weather which travels from west to east along a path determined by the jet stream pattern. This is a general description of what has long been known as a cyclone, a traveling weather disturbance with rotating winds.

WEATHER IN A CYCLONE

The description of the jet stream and cyclone above applies to events occurring at altitudes of about 17,000 meters. The resultant surface weather forms at much lower altitudes. It is most conveniently described in terms of air masses and fronts. In the lower part of the atmosphere large areas tend to be covered with air having nearly uniform temperature and moisture content. These are called *air masses*. Air mass characteristics depend on the land below and distance from the equator—warm and dry over desert areas, warm and moist over southern seas, cold and dry over the northern continent, and so on. The sharp boundaries between air masses are *fronts*. The fronts in the atmosphere are almost invariably in motion. When warm air is advancing on cold air the front is called a *warm front;* when cold air is advancing on warm, it is a *cold front*. The formation of a cyclone involves a perturbation of the cold air/warm air interface so that a warm front is formed ahead of the cyclone and a cold front behind (Fig. 1.9). As the cyclone develops in intensity the boundary between warm and cold air develops a progressively greater perturbation, as illustrated in Figure 1.10a–c, while weather conditions at the disturbance deteriorate. This is a cyclonic storm. It is carried along in a generally southwest to northeast track by the jet stream. Particularly in the winter months, several such storms pass over North America each week.

Before the characteristic weather pattern of the cyclone can be described it will be necessary to examine the weather-making processes that occur in the lower atmosphere.

TEMPERATURE AND PRESSURE

Ordinarily, as shown in Figure 1.11, the temperature in the atmosphere decreases above ground level until a minimum temperature is reached at about 10,000 meters. The zone in which temperature decreases with height is called the *troposphere* (the

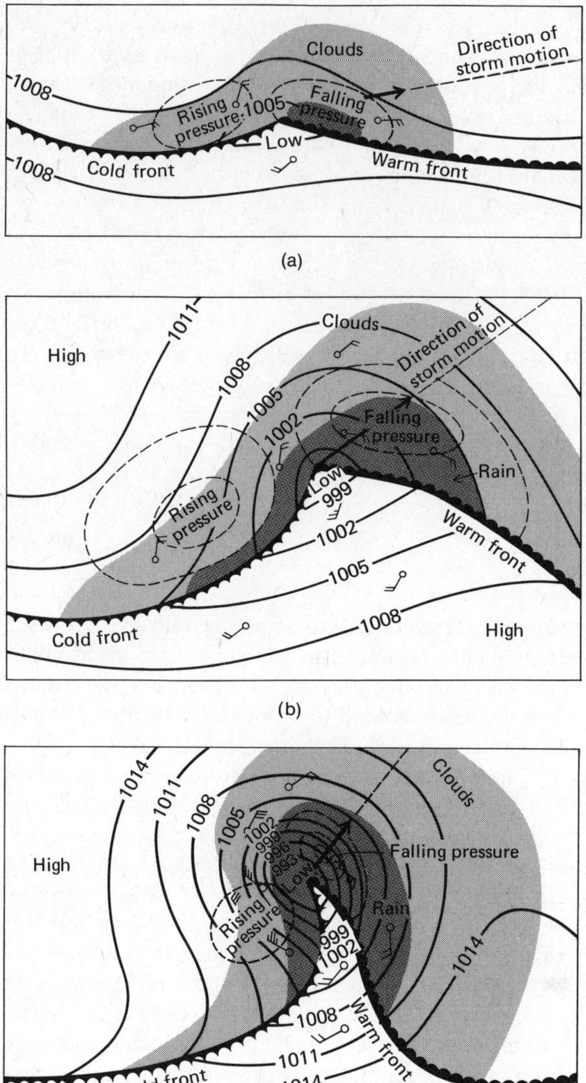

Figure 1.10 Successive changes in the development of a cyclone. Lines of constant air pressure, isobars, are shown in each diagram and are identified by the pressure in millibars (1008, 1005, and so forth). Wind arrows show the direction and, by the number of tail feathers, intensity of the wind. Fig. 1.10(a) corresponds to a weak storm; Fig. 1.10(c) corresponds to an intense storm. Compare the wind pattern with that in Figure 1.9. After Bowditch, *American Practical Navigator,* through courtesy U.S. Naval Hydrographic Office, 1958.

mixing layer); above it is the *stratosphere*. This temperature distribution results because the earth's surface acts as the principal heat source for the atmosphere, heating it from below, while additional heat flows down from high in the stratosphere where the incoming ultraviolet radiation is absorbed. The rate of temperature decrease with altitude in the troposphere, called the *temperature lapse rate,* is about 0.65°C per 100 meters (3.5°F per 100 ft). Sometimes *temperature inversions* occur near the surface, as in Figure 1.12. At the inversion, temperature *increases* with altitude over a short interval. Inversions usually occur on clear summer nights; when there is no cloud cover the earth's surface temperature drops rapidly at night as heat

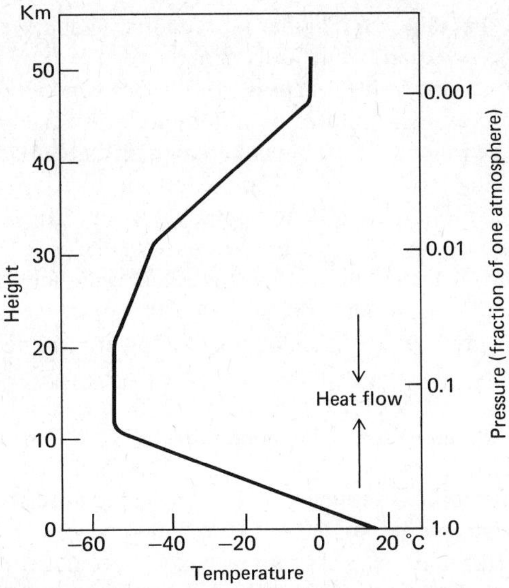

Figure 1.11 Variation of temperature with altitude in the atmosphere.

is radiated away. The bottom layer of the atmosphere near the cool ground is cooled by conduction and convection while the air above is not able to lose much heat in the same time. By morning the temperature profile is like that in Figure 1.12.

The weight of the air high in the atmosphere must be supported by the air lower down. Consequently, air pressure increases downwards through the atmosphere. Atmospheric pressure at the earth's surface is about 1 bar or 14.7 pounds per square inch, sufficient to support a column of water about 10 meters (32 ft) high or a column of mercury of 76 cm (about 30 in.). This fact is utilized in the construction of the mercury barometer for the measurement of atmospheric pressure. Because there are small but characteristic changes in air pressure during the passage of a cyclone, the barometer is a useful instrument in weather forecasting. In an intense storm surface pressure may drop by about 6 percent.

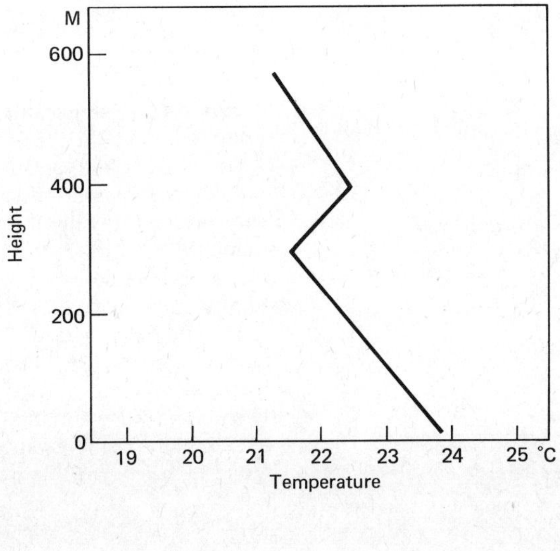

Figure 1.12 The variation of temperature with altitude in the troposphere when a temperature inversion is present. The inversion in this case is between 300 and 400 m.

If a volume of air at ground level is taken to a higher altitude it expands, because of the lower pressure. Now when air is expanded, it cools (feel the temperature of the air blown off by a pneumatic tire when the valve is removed). If there is no exchange of heat with the surroundings, the cooling is said to be *adiabatic*. Air raised in the atmosphere cools adiabatically at the rate of 1°C per 100 meters of altitude (5.5°F per 1000 ft). Because the adiabatic cooling rate is greater than the temperature lapse rate, the lower atmosphere is ordinarily *stable*—there is no spontaneous exchange of air between high and low altitudes. The cause of this stability is easily seen with the aid of the following thought experiment: Take a volume of air to a slightly higher altitude and release it. If the air brought to the higher altitude has a density greater than that of the surrounding air (because it is cooler), it will sink back to a lower altitude. This means that the atmosphere is stable because it resists a perturbation applied to it. If, however, the density of the test volume of air is less than that of the surrounding air, the atmosphere is unstable: Any small disturbance which raises a volume of air will result in that air continuing to rise.

Now let us apply the concept of stability to rising air which is heated at the earth's surface. Suppose that because of strong sunshine on a summer day, air near the ground is warmed a few degrees. Warm air expands and so has a lower density than cold air at the same pressure. Hence the heated air near the ground becomes buoyant and begins to rise. The construction in Figure 1.13 shows how the height to which it can rise is determined. The warmed air cools at the adiabatic rate (gray line) as it rises, hence its temperature drops more rapidly than that of the surrounding atmosphere (black line). At the intersection point of the two cooling curves the buoyant

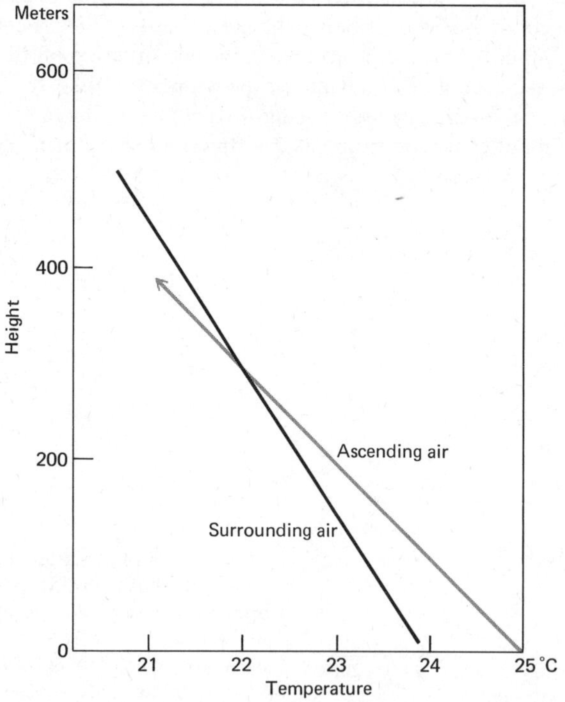

Figure 1.13 Temperature-altitude curves for the atmosphere (black line) and for a volume of air warmed at the surface (gray line). Air heated at the surface is unstable and rises. Its temperature drops more rapidly than that of the surrounding air. The buoyant force becomes zero when the two curves cross and the ascent of the rising air soon stops.

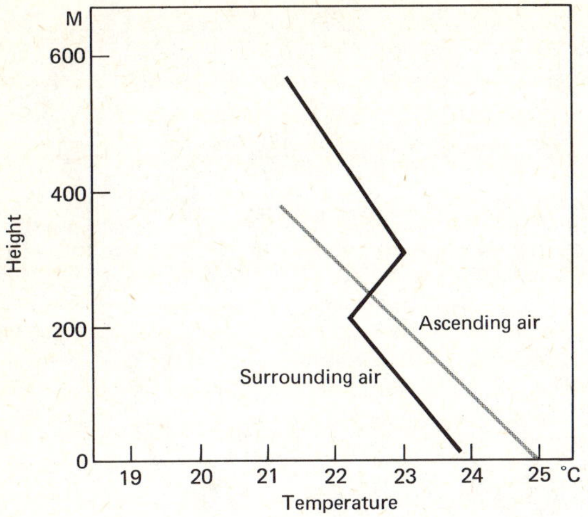

Figure 1.14 A temperature inversion, such as is shown in the black temperature vs. altitude curve, blocks the rise of heated air whose temperature follows the gray line as it rises. The inversion may be at low altitude and so act as a trap for waste gases emitted below.

force on the rising air becomes zero because its density is now the same as that of the surrounding air. This air will not rise much higher. The same argument applies to the emission of stack gases from a chimney. The height to which the warm gas emitted from the stack, and, hence, the height to which the entrained products of combustion can be carried, is determined by a diagram like that of Figure 1.13. When there is a temperature inversion present a much lower bound on the height to which warm gas can rise is likely to be established, as shown in Figure 1.14. An inversion at a low altitude over a city traps all stack and exhaust emissions below it. This can result in severe deterioration of air quality at the surface. There have been three occasions in recent years when an inversion at low altitude over a city trapped enough waste gases to make the city air lethal for many persons: in 1930 in the Meuse Valley, Belgium, in October 1948 at Donora, Pennsylvania, and in December 1952 in London. In each case stack gases containing oxides of sulfur formed by burning coal were trapped below an inversion. In London the problem has been solved by banning the use of sulfur-containing fuels; there has been a substantial increase in the number of sunny days in the wintertime. Temperature inversion in the vicinity of Los Angeles, California, trap automobile exhaust emissions and are responsible for the frequent occurrence of smog in this locality. Again, a solution must be found in the control of the waste emissions; there is no way of controlling the inversions.

MOISTURE

The amount of water vapor present in air determines its *humidity*. Absolute humidity may be expressed as the number of grams of water vapor per gram of air. For most purposes in meteorology the *relative humidity* is of more interest. There is an upper bound to how much water vapor the air can hold. When this bound is reached the air is *saturated*. Any water vapor present in excess of saturation will be precipitated as a condensate, such as fog, rain, snow or sleet. Relative humidity is the amount of moisture in the air relative to the amount it can hold at its temperature. Fifty percent relative humidity means that the air is 50 percent saturated. The relative humidity deter-

Figure 1.15 Fog formed over the cold waters of the Pacific Ocean advancing through the Golden Gate. National Weather Service Photo.

mines evaporation rates—as the relative humidity increases, evaporation rates decrease. For this reason evaporation cooling (as from the skin) becomes ineffective during damp weather.

The saturation water vapor content of air depends on temperature; the higher the temperature the greater the amount of water vapor that can be contained without precipitation. If warm, moist air is cooled, the relative humidity will increase. When it reaches 100 percent the *dew point* temperature has been reached and moisture begins to condense (as dew) on any solid surface in contact with the air. Near the earth's surface, relative humidity is likely to be high because of evaporation of water from the sea and transpiration by plants on land. If surface air is carried upwards it cools, the relative humidity increases and eventually saturation occurs. Moisture then condenses out of the air as small droplets and forms a cloud. Cooling air at the earth's surface can also result in condensation. In this case *fog* is formed, as illustrated in Figure 1.15. Fog is simply a cloud at ground level.

There are two general cloud types, stratus and cumulus. If warm, moist, stable air is forced upwards because of a large scale air circulation pattern, such as that associated with a cyclone, condensation occurs in a thin sheet spread over a wide area. This thin, flat stratified cloud is called *stratus*. If the upwards motion of air is the result of instability produced by surface heating, the air rises in irregular columns within which strong buoyant forces drive the air to high altitude. The resultant clouds have a strong vertical development and often show evidence of internal turbulence.

Table 1.1 Cloud Classification

	3000 Meters (Low)	*4000 Meters (Middle)*	*7500 Meters (High)*
Stratus type	Stratus	Altostratus	Cirrostratus
Cumulus type	Cumulus	Altocumulus	Cirrocumulus

They are *cumulus* clouds. The basic classification of clouds is based on cloud type and altitude of formation, as shown in Table 1.1.

Clouds which are actively producing precipitation are called *nimbus*. From *nimbostratus* a steady rain falls but *cumulonimbus* produces showers (which may, however, be very intense). The principal cloud types are illustrated in Figure 1.16.

THE CYCLONIC STORM

Let us re-examine Figure 1.10, showing the characteristics of a cyclonic storm in its stages of development. The pattern of surface air pressure is shown by the *isobars* (lines of constant pressure). In these drawings the pressure at each isobar is indicated in units of millibars (1000 millibars = 1 bar). Pressure is lowest at the eye of the storm, into which the surrounding surface air is flowing. The circulation pattern is shown by the wind arrows (which point in the direction in which the wind blows). Hatch marks on the arrows indicate wind velocity. Warm, moist air entering the eye is carried upwards where moisture condenses out to form an extensive blanket of stratus clouds. The approach of a cyclonic storm is foretold by a steadily falling barometer, shift of the wind to east or southeast, and the successive advance of cirrostratus, altostratus, stratus, and nimbostratus clouds from the southwest. The arrival of cirrostratus is often revealed by a halo about the sun or moon. At this stage the storm center is about 1000 km away; this is often the first sign of its advance. The surface air circulation about the cyclone or "low" as it is sometimes called, is shown by the wind arrows of Figure 1.10 to be a counterclockwise spiral in towards the low pressure. At the storm center there is no surface wind at all. When a storm passes directly over an observer, the easterly wind of the storm's approach stops quite suddenly, calm prevails for perhaps 10 to 30 minutes, whereupon the wind commences from the *opposite direction* with great intensity. This sudden and often unexpected wind shift has been the demise of many vessels, particularly sailing vessels, which have successfully weathered the first half of the storm.

Under the influence of the jet stream above, the cyclone is carried along a generally northeasterly course (shown by the broad arrows in figures). To the southeast of the storm center the total wind speed is the sum of the circulation speed and forward velocity of the storm center. On the west side the wind speed is reduced by the speed of forward motion of the storm center. The southeast side of a cyclonic storm is, therefore, always more intense; it is called the *dangerous semicircle,* the northwest side, the *navigable semicircle.* A ship at sea maneuvering to avoid an intense storm must always steam towards the northwest side of the cyclone. This is the *law of storms,* well known to mariners long before cyclonic circulation was discovered.

Cross sections through the warm and cold fronts of the cyclone are shown in Figure 1.17. At the warm front warm air rides up and over the cold air below. The

(a)

(b)

(c)

Figure 1.16 The principal cloud types: (a) Cumulus; (b) Stratus; (c) Altocumulus.

(d)

(e)

(f)

Figure 1.16 The principal cloud types (*cont.*); (d) Altostratus; (e) Cirrocumulus (mackerel sky); (f) Cirrostratus (mare's tails. (a) and (d) courtesy of the American Museum of National History; (e) and (f) courtesy Ward's National Science Establishment.

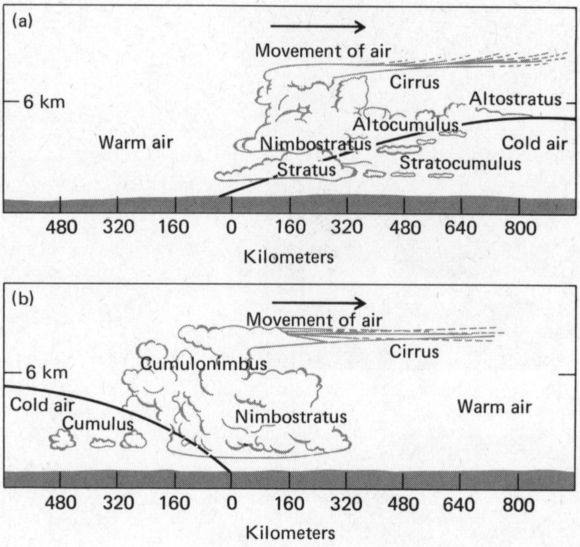

Figure 1.17 Cross section through a warm front (a) and a cold front (b). Courtesy National Weather Service.

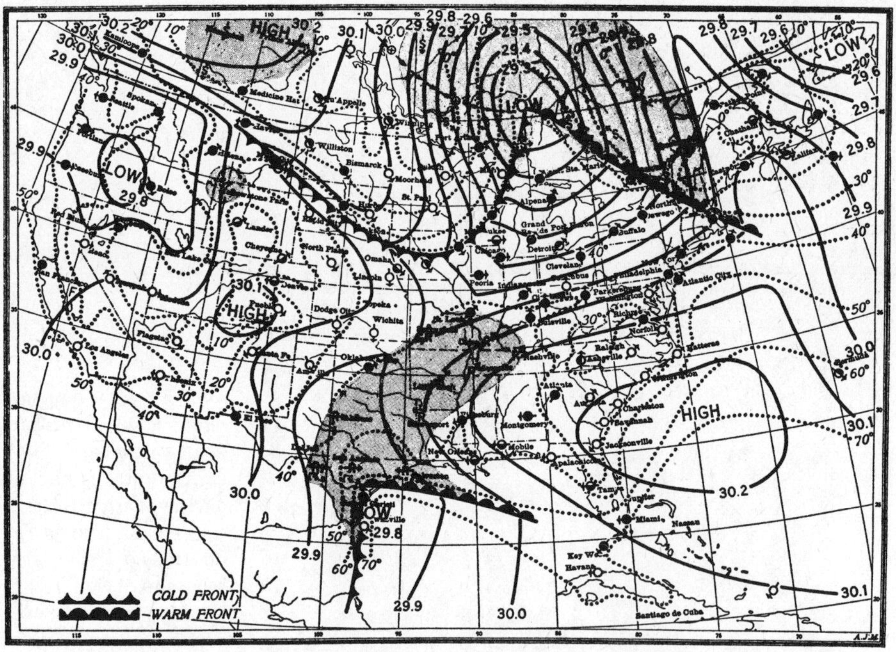

Figure 1.18 A weather map for the United States showing a well-developed storm system to the north of the Great Lakes. Isobars are solid lines with pressure in inches of mercury. Isobars are dotted with temperature in °F. Station symbols indicate cloud cover and wind direction. Symbols for fronts are as in Figure 1.9. National Weather Service, U.S. Department of Commerce.

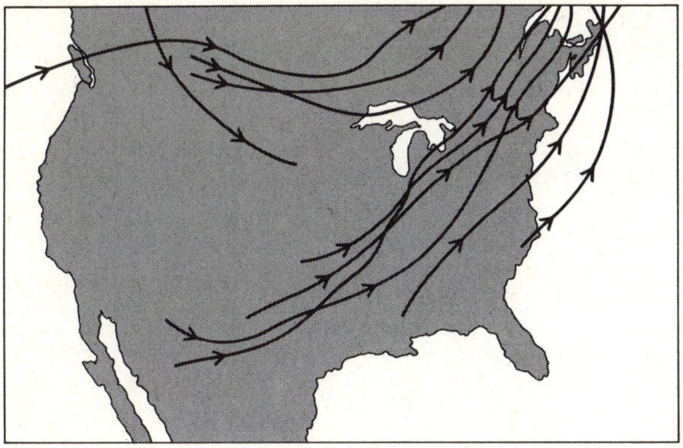

Figure 1.19 Typical set of storm tracks over the United States during the month of January. From *Introduction to the Atmosphere* by Riehl. Copyright 1965 by McGraw-Hill. Used with permission of McGraw-Hill Book Company.

slope of the boundary (its rise for a given distance along the ground) between the warm and cold air is low, 1/100 to 1/300, so cirrostratus clouds may be seen up to 1000 km ahead of the place where the boundary between cold and warm air reaches the ground. At the cold front cool air pushes in under the warm air ahead. The front is much steeper, its slope may be as sharp as 1/25, and its passage usually brings a rapid change in weather. Usually the cold front is marked by cumulus clouds and often there is thunderstorm activity. As the front passes the wind shifts to the northwest, the temperature falls and the air clears.

Measurements of air temperature and pressure and observations of wind and cloud conditions are now made at numerous land stations and by many ships at sea on a regular schedule. When this data is collected and plotted a *weather map* may be produced. A typical map is shown in Figure 1.18. Station symbols show the amount of cloud cover and wind characteristics. Isobars and fronts are shown by the symbols already used in Figure 1.10. Note that each of the drawings in Figure 1.10 is in fact a small section of a weather map. Comparison of weather maps made for successive days permits determination of the tracks followed by cyclones. The tracks of cyclones moving over the United States during a typical January are shown in Figure 1.19. Many of these storms pass out to sea through the St. Lawrence Valley. Others move up the east coast of the United States where they generate the northeasters which make New England winters so notorious. The wind pattern and track of one of the most intense of these storms, that of March 7, 1962, is shown in Figure 1.20. The northeast winds driving across the open Atlantic are responsible for combination of high water and high waves which, in this storm, caused extensive loss of property along the Atlantic Coast. Photographed from very high altitude, as from a satellite, the complete cloud pattern of a cyclone can be seen. Figure 1.21 is an example.

Most storms in the northern hemisphere are caused by cyclones spawned by the prevailing westerlies. Some cyclones originate, by mechanisms not yet very well known, in the tropical oceans. In the Atlantic such storms are known as hurricanes, in the Pacific as typhoons. Their storm track is curved but generally northerly in trend. Hurricanes and typhoons occur infrequently (compared to other cyclonic storms) but may be both very large in area and very intense.

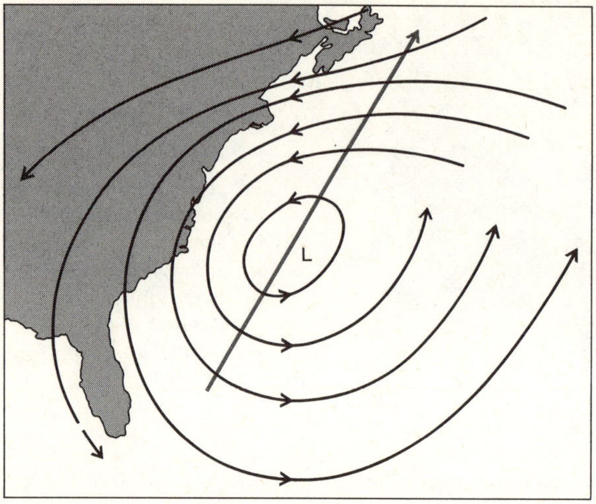

Figure 1.20 Circulation pattern and storm track (gray line) for the severe north-eastern gale of March 1962. Winds along the Atlantic coast approached hurricane force during this storm. From *Introduction to the Atmosphere* by Riehl. Copyright 1965 by McGraw-Hill. Used with permission of McGraw-Hill Book Company.

Figure 1.21 Cyclonic disturbance over the Pacific Ocean about 1200 miles north of Hawaii as seen from Apollo 9. NASA photograph.

WEATHER FORECASTS

The energy of even a modest cyclone is so enormous compared to the energies that can be generated by processes controlled by man that the prospects for being able to modify or deflect cyclonic storms seem remote. Accurate forecasts, particularly of intense storms, are therefore very desirable. There are a number of possible approaches to the problem of forecasting weather. The oldest, and still most effective, is based on experience. A definite pattern of weather changes heralds the approach of a cyclone. If these are recognized, appropriate warnings can be issued. If through reports from a network of observatories the general pattern of weather is known across the nation, a fair guess can be made of the future path of the storm. But as the track chart in Figure 1.19 shows, storm tracks are not perfectly regular. Detailed information about the jet stream flow that controls the storm path is not usually available to the forecaster. Inevitably, there is uncertainty as to the exact course the storm will follow. Unfortunately, only a small change in track can mean the difference, for example, between heavy snow and heavy rain in New York City. But the weather forecaster is expected to supply just this sort of detail. While the arrival of a storm may be predicted with good confidence several days in advance, it may be impossible to predict important details about it with satisfactory reliability.

A second method of forecasting, which is essentially an elaboration of the first, relies on a catalog of past weather events. As complete a picture as possible of the current state of the atmosphere is assembled. This is compared to the previous recorded states. If in the record a past state can be found which exactly matches the present state, then the record of events which followed that past state is the present forecast. There are two problems with this method: One is to know in what detail the condition of the atmosphere must be described in order to know that two states are identical. The second is the size of the catalog needed so that there is a good chance that a matching past state can be found. Unfortunately, it now appears that an adequate catalog could only be accumulated by detailed observations made for at least 1000 years. At present, therefore, this method of *analogs* is not too useful.

The third and most direct method of forecasting is by numerical calculation of the future motion of the atmosphere. In principle it is the same method as that used in celestial mechanics to predict the future positions of the planets. It was shown many years ago by Lagrange that if the position and velocity of every body in a planetary system is known at some instant of time, the position of every body at all future times is determined. In the solar system the computational difficulties are great, but accurate predictions of future events, such as eclipses, can be made with a good degree of accuracy. In applying the method to the weather there are difficulties in both principle and practice. First, the governing equations are not exactly known because the physics of some important processes is not fully understood. The process that collects the finely dispersed water droplets in a cloud into raindrops is one. Next, the equations, so far as known, are nonlinear (effects not proportional to causes) and so are difficult to solve. Approximate numerical methods of solution must be used. Finally, the atmosphere is a continuum, but for numerical solution of the equations it must be represented by numerical values determined on a grid of points. There are real limits both in observation and computation on how fine this grid of points can be. Motions that are small compared to the grid size will be lost in the solutions of the equations, but these motions may be important in the future development of large-scale weather systems. Numerical forecasts are now being made by the U.S. Weather Service and are successfully used to supplement the forecasts made by traditional methods. However, the costs of acquiring the more detailed observations and doing the finer scale calculations required for a major increase in forecast accuracy are so great that an improvement of more than a few days in the time for which reliable forecasts can be made is not likely to be attained. The vagaries of the weather will remain a topic of conversation for years to come.

CLIMATE—IS TECHNOLOGY CHANGING IT?

The long-term weather pattern is climate. There is evidence that human activities in industrialized society are causing climatic change now and that they may cause much more severe changes in the future. Temperature records show that in about 1880 the average surface temperature of the earth began to rise; by 1940 a much more rapid decrease in temperature began. The decrease since then has been about 0.2°C. These seem like small changes but a decrease in average temperature of only 4°C is enough to start an ice age. The change of a few tenths of a degree significantly alters the amount of ice cover on the polar seas and the distribution of important food fish species. There are a number of ways in which man's activities could be causing these temperature changes.

Since 1900 the average CO_2 content of the atmosphere has increased from about 290 ppm (parts per million) to about 330 ppm. The increase is due to the discharge into the atmosphere of the products of combustion of fossil fuels such as coal and petroleum. Carbon dioxide in the atmosphere is responsible for the greenhouse effect, the blanketing of the earth by a layer opaque to the earth's thermal radiation which keeps the surface temperature higher than it would otherwise be. Calculations (which can only be done approximately) show that the increase in CO_2 could raise the surface temperature by about 0.2°C, which is at least of the order of the increase observed. The decrease in temperature since 1940 must clearly be due to some other cause.

In recent decades there has been a large increase in the amount of dispersed particle matter in the atmosphere, as much as an 80 percent increase in the northern hemisphere during the last decade alone. Sources of these fine particles include dust stirred up by cultivation of the land but more importantly, particles resulting from combustion of fuels in furnaces and propulsion machinery of all kinds. The direct effect of these particles is to reduce the transparency of the air; the amount of radiation reaching the earth's surface is reduced and the average surface temperature falls. There is also an important indirect effect. Particles suspended in the air are nuclei on which moisture can condense. Associated with the air pollution found in urban areas is an increase in the amount of rainy and foggy weather. These changes are well documented for the great industrial cities of Europe and North America. It also seems likely, though not yet proven, that there will result an increase in the average cloud cover world-wide. A 1 percent increase in average cloud cover will decrease the average surface temperature by 0.8°C. There is little question that the discharge of particulate matter into the atmosphere can produce significant surface temperature changes. Proof that the observed temperature decrease actually is so caused is still lacking. When and if such proof is found it is clear that we will face an urgent problem since at present the rate of discharge of particulate wastes into the atmosphere is accelerating.

The air pollution problem is a good example of how difficult it is to recognize fully and to apportion fairly the costs of a technological development. Industrial activity in Europe and North America may be making life more difficult for the residents of the more northern countries, clearly a cost that is not fairly assessed against those who benefit from industrialization. Other examples are easy to find: supersonic transport aircraft benefit a few travelers but the cost is, in part, borne by the much larger number of persons who must live with its noise and, perhaps, the climatic changes caused by discharge of its exhaust products high in the atmosphere. Conventional economics and marketplace decision-making are inadequate to deal with this class of problem.

Further Readings

BOWDITCH, N., *American Practical Navigator.* U.S. Navy Hydrographic Office, 1958. [Chapters 37, 38 and 39]

CRITCHFIELD, H. J., *General Climatology.* Englewood Cliffs, N.J.: Prentice-Hall 1966.

MCINTOSH, D. H., THOM, A. S., and SANDERS, V. T., *Essentials of Meteorology.* New York: Springer-Verlag, 1969.

PETTERSON, SVERRE, *Introduction to Meteorology.* New York: McGraw-Hill, 1969.

RIEHL, H., *Introduction to the Atmosphere.* New York: McGraw-Hill, 1965.

SUTCLIFFE, R. C., *Weather and Climate.* New York: W. W. Norton, 1966.

TRICKER, R. A. R., *The Science of the Clouds.* New York: Elsevier, 1970.

——— *Inadvertent Climate Modification.* Cambridge (Mass.): MIT Press, 1971.

2 Exploring the Earth's Surface

Travel over dry land is a relatively easy matter; even the most remote outposts can now be quickly reached. Exploration of the continents is essentially complete. But more than two-thirds of the earth's surface is covered by ocean. Extending over thousands of kilometers, with an average depth of only a few kilometers, the oceans are a thin veneer of water on the earth. This layer of water makes it very difficult to study the land beneath the sea. Indeed, the surface of the moon is better known to us than is the sea floor of our own planet. This chapter describes the science involved in exploring the earth's surface, with special emphasis on the oceans. Chapters 5 and 8 return to the study of the face of the land above the sea.

MAPS AND NAVIGATION

Searching out the geography of the oceans began when boats were first built, reached its zenith in the great voyages of discovery of the 16th, 17th, and 18th centuries, and has only in recent years been brought near to completion. Central to voyages of exploration has been the problem of finding the position of a ship at sea (where there are no fixed points of reference); indeed, the needs of navigation were one of the first motivations for the study of the physics of the earth. Refinement of the methods of marine navigation has led to the science of *geodesy,* concerned with the determination of the size and shape of the earth, and to methods of air and space navigation.

The location of any place on the surface of the earth is specified by its *latitude* and *longitude*, the parameters of the spherical coordinate system shown in Figure 2.1. Latitude is the angular distance above (North latitude) or below (South latitude) the equator. Curves of constant latitude are called parallels, of constant longitude, meridians. The longitude of a place is the angular distance eastward (East longitude) or westward (West longitude) from the meridian of Greenwich, England.

Latitude and longitude are specified in angular measure, but for many practical purposes the distances between places in kilometers or miles is required. To know the conversion factor from angular to linear measure, the radius of the earth must be known; its determination was one of the first geophysical problems to be solved. The

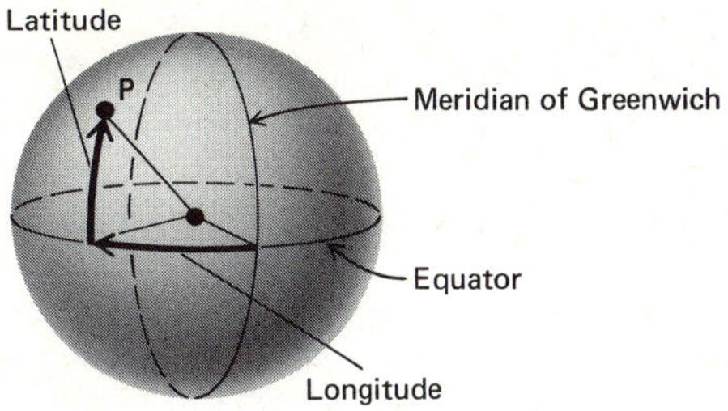

Figure 2.1 The location of the place P is specified by its latitude and longitude, the angles shown in the drawing.

method used by Eratosthenes in the third century B.C. illustrates the procedure. Eratosthenes discovered that a certain well in Syene (near Aswan, Egypt) was lighted throughout its depth by the sun only on the first day of summer; that is, the sun was directly over the well on this one day of the year. The well was (in modern measure) 800 kilometers south of Alexandria. On the first day of summer Eratosthenes measured at Alexandria the length of the shadow cast by a vertical pole of known length. He used the lengths of the pole and its shadow to find the angle γ in Figure 2.2. It is 7.5°. In modern measure 7.5° of latitude corresponded to 800 kilometers, so Eratosthenes calculated that 360° corresponded to 38600 km. The modern value for the length of the equator is 40000 km. This makes a nautical mile, the length of a degree of arc at the equator, equal to 1852 meters. (Because Columbus used much too small a value for the length of the nautical mile, he badly underestimated the distance he would have to sail west to reach India from Spain. It was his luck that America happened to be at about his point of no return.)

Locations of places are specified by latitude and longitude; distances between them can be calculated by spherical trigonometry. But for many purposes a graphical

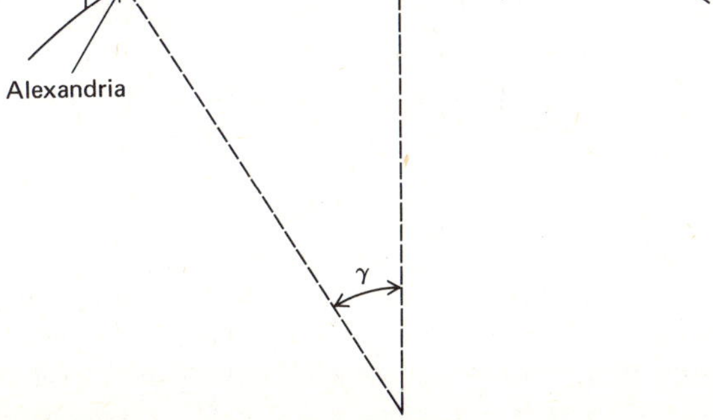

Figure 2.2 On the first day of summer, the sun passes directly over the well at Syene; at Alexandria, 800 km away, a vertical pole casts a shadow at an angle $\gamma = 7.5°$. From these data the circumference of the earth can be calculated.

rather than an analytic representation of geography is most useful. A globe gives an accurate representation of the relative positions of points on the earth's surface but is awkward to use when a large scale is required. A more convenient map or chart can be made by projecting the features marked on a globe onto a flat surface. Unfortunately, there is no way of representing features of a spherical surface on a plane without some type of distortion. Map projections can be devised to accurately represent shapes, areas, distances or directions, but no projection is accurate for more than two of these. For navigation over distances of hundreds of miles or less a projection in which relative directions are accurately represented is most useful. The Mercator projection has this property; a straight line connecting any two points has a constant bearing (angle relative to north), so that a compass course can be read directly from a Mercator chart. The first step in the construction of a Mercator chart is to project points from the globe onto a cylinder set up tangent to the equator (Fig. 2.3). When unrolled the cylinder becomes a map in *cylindrical projection.* To convert this to Mercator projection the distances of points from the equator must be reduced by amounts which increase with latitude. (The amount of this reduction is found by an intricate calculation first completed by Edmond Wright after Mercator's concept.) The main disadvantage of the Mercator chart is the great distortion of areas at high latitudes; it is not a good choice of projection to use when, for example, it is desired to compare the areas of countries which are at different latitudes. Nor does it show as

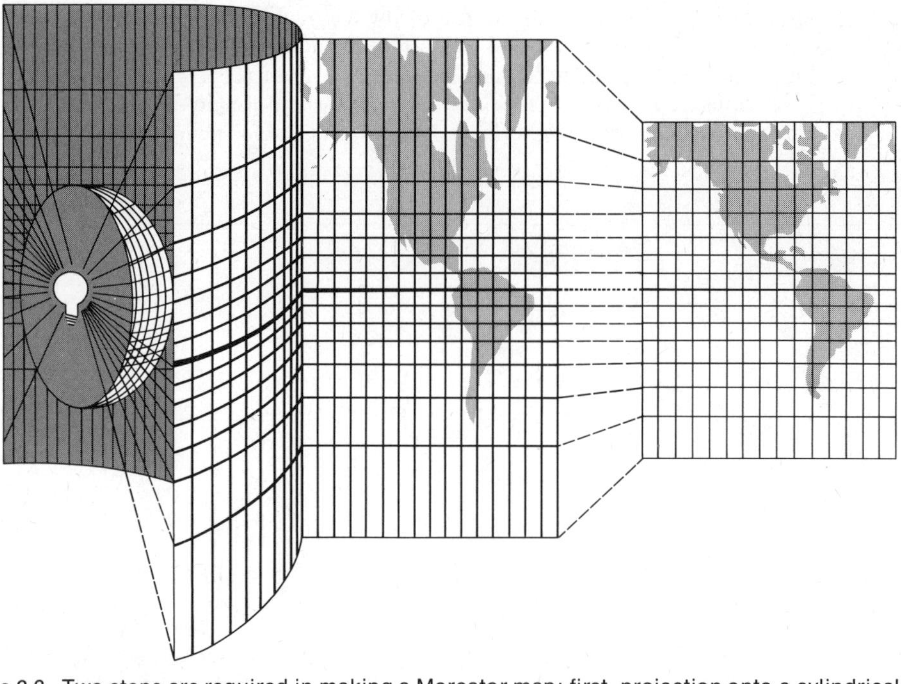

Figure 2.3 Two steps are required in making a Mercator map: first, projection onto a cylindrical surface; second, adjustment of the spacing of the parallels. From David Greenwood, *Mapping,* University of Chicago Press, 1964. Copyright 1964. Adapted with permission of University of Chicago Press.

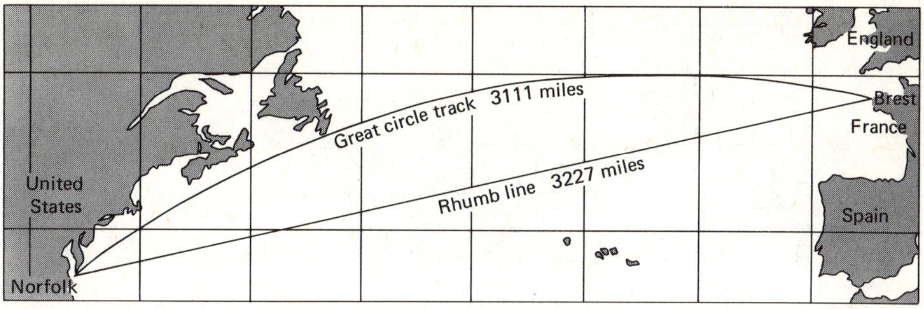

Figure 2.4 The shortest (great circle) route between the two ports is not a straight line on a Mercator chart. The *rhumb* line is a track having a fixed compass course.

a straight line the shortest (great circle) route between two distant places, illustrated in Figure 2.4.

A great variety of map projections, each having a different combination of desirable and disadvantageous features, has been devised. Some of these are compared in Figure 2.5. Gall's stereographic projection eliminates much of the distortion of area found in the Mercator chart. The Mollweide projection shows correct relative areas. On the azimuthal equidistant projection the true distance and direction from one place (London in the example) to any other place can be read. Figure 2.5d is a chart of the North Atlantic Ocean on a gnomonic projection. In this projection great circles plot as straight lines. It is useful when the most direct air or sea route between ports is to be found. Clearly, a map projection should be chosen relative to the intended use; largely for historical reasons the Mercator projection tends to be much overworked by map makers.

In order to construct a chart or globe the latitude and longitude of the places to be mapped must be determined by observations in the field. The primary method of establishing the position coordinates of a place on land or at sea is by astronomical observations, usually of stars or planets but, most recently, of artificial satellites also. The principles involved can be illustrated by one of the simplest observational techniques used, the determination of latitude from a meridian altitude.

Because of the rotation of the earth, any celestial body—the sun, a star or planet or the moon—rises in the east, traverses an arc across the sky and sets in the west. The *altitude* of a celestial body at any time is its angle of elevation above the horizon. The altitude of any body is greatest when it passes the meridian of the observer. The instant of meridian passage can be determined by observing the altitude with a sextant (see Appendix to this chapter) or telescope over a period of time: At the instant the body stops rising and begins to set, it is on the observer's meridian. This is illustrated by the altitude data for the sun plotted in Figure 2.6. (When the sun is on the meridian, it is noon in local time. A meridian altitude of the sun taken on board ship is known as a *noon sight* and may be used to reset the ship's clock to allow for her eastward or westward progress.)

The method of calculating latitude from the observed meridian altitude can be worked out from Figure 2.7, which is drawn in the plane of the observer's meridian. The observer at O measures the altitude, α, of the celestial body above his local horizon, h—h′. The declination of the body, d, is its elevation above the equator. (The declinations of all celestial bodies useful in navigation are known from observations

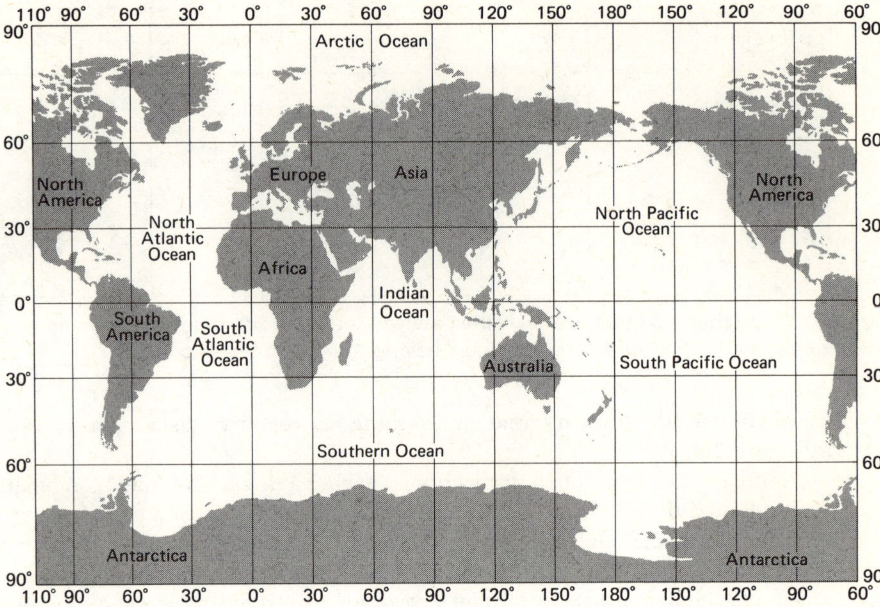

Figure 2.5 Comparison of map projections: (a) Gall's stereographic eliminates most of the distortion of area found in the Mercator projection.

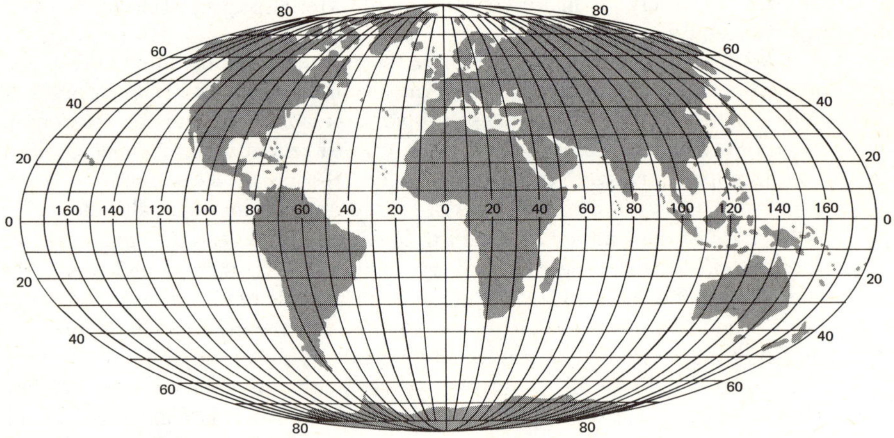

Figure 2.5 Comparison of map projections (*cont.*): (b) Mollweide's projection shows all areas in true proportion.

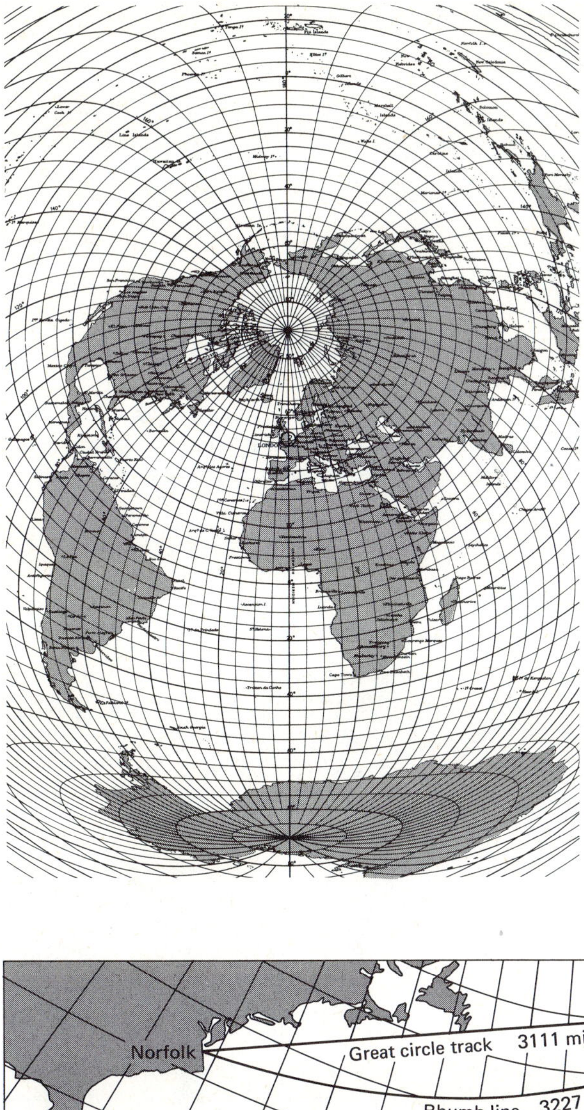

Figure 2.5 Comparison of map projections (*cont.*): (c) On the azimuthal equidistant the true direction and distance to any place from the central point (London in this case) can be read directly from the chart. Reproduced from BA chart 5085 with the sanction of the Controller, HM Stationery Office and the Hydrographer of the Navy.

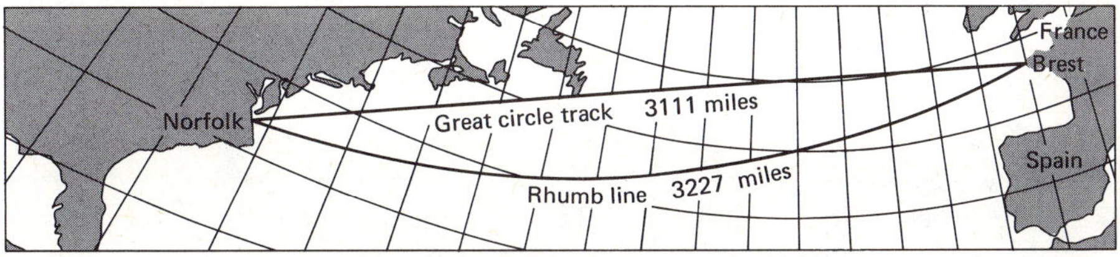

Figure 2.5 Comparison of map projections (*cont.*): (d) gnomonic projection.

made at astronomical observatories and are tabulated in almanacs.) It follows from the figure that the latitude is

$$L = d + \beta = d + (90 - \alpha) \tag{2.1}$$

Longitude is measured along the direction of rotation of the earth. Consequently, the determination of longitude by astronomical observations requires accurate

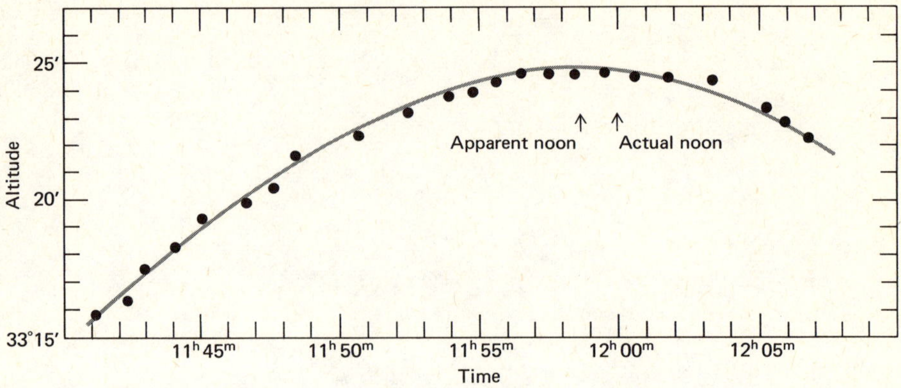

Figure 2.6 A series of altitude observations of the sun taken on a ship with a marine sextant. By averaging the observations (constructing a smooth curve through the points) the altitude at noon can be determined to within ±0.1′ and the time of noon, to about one minute. From these observations the observer's latitude can be calculated as in Figure 2.7.

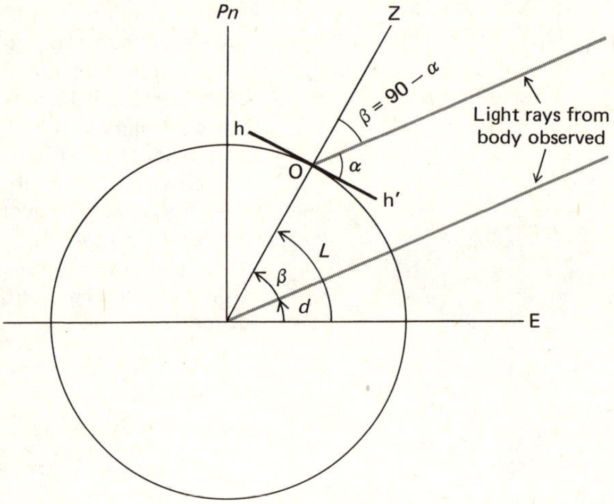

Figure 2.7 The latitude of the observer at O is found from the measured altitude, α, by this construction. The declination of the body observed is *d*; hh′ is the horizon seen by the observer, Z is the point (the zenith) directly over the observer, and E, the plane of the equator.

knowledge of time. If, for example, it is known that a certain star is on the meridian of Greenwich at time t_1, and crosses the meridian of the place of interest at time t_2, then the longitude of that place measured westward of Greenwich is $(t_2 - t_1)/p$, where p is the time required for the earth to rotate 1° ($p = 24$ hr/360°). Until late in the 18th century accurate timekeepers suitable for use on ships were not available and navigators were unable to determine their longitude by any convenient method of astronomical observation. The common method of sailing was to proceed to the latitude of the port to be made (latitude was determined by noon sights for which accurate time is not required), then run at constant latitude for a landfall (which, hopefully, did not occur during thick weather). Harrison's invention of the chronometer (1759) made possible accurate timekeeping on ships and the accurate determination of the longitudes of places visited by explorers. (Captain Bligh of the *Bounty* made some of the first determinations in the Pacific. The technical and practical difficulties encountered by the early explorers are recounted in the book by Ritchie listed at the end of the chapter.) Since radio time signals have become available, longitude observations comparable in accuracy to latitude have been possible.

Celestial navigation is capable of fixing a position to within a few tenths of a mile for a ship at sea, and to within a few meters on land. But it has important practical limitations for regular use at sea: Since to fix a position two or more bodies must be observed, positions can only be determined at dawn and dusk when both the horizon and a number of stars can be seen. The difficulties which arise during a protracted spell of bad weather are obvious. The limitations of celestial navigation are likely to be particularly vexing on an oceanographic research ship where positions must be obtained frequently and accurately if the oceanographic observations being made are to be of any value.

A number of methods of position finding which rely on electronic methods are now in use. *Inertial navigation* is essentially very sophisticated *dead reckoning:* The departure point of the vessel is known. Accelerometers are used to measure every acceleration of the ship or aircraft after departure. Integration of the accelerations gives the velocity at every instant of time; integration of the velocities, the position. Obviously a sophisticated measuring and computing system is required, the expense of which can usually be justified only for the most pressing military missions.

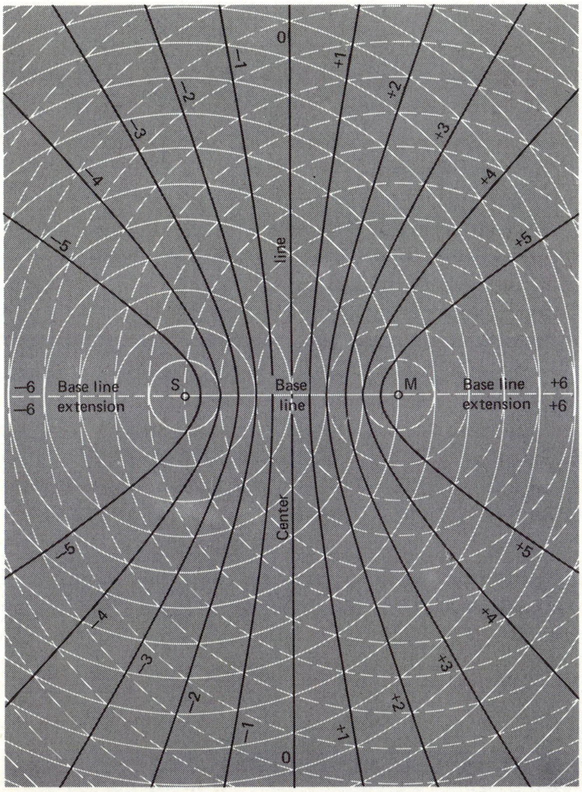

Figure 2.8 A Loran station consists of master (M) and slave (S) transmitters. Radio signals are sent simultaneously from M and S and at successive later times reach positions on the circles. On each black curve there is a constant time difference between receipt of the signals from M and S. Measurement of the time difference fixes the observer on one of the curves. Measurement on a second station locates his position (where the two time difference curves cross). From Bowditch, *American Practical Navigator,* through courtesy U.S. Naval Oceanographic Office.

Hyperbolic navigation relies on the reception of timed radio signals from two or more pairs of stations. The widely used "Loran" (LOng RAnge Navigation) system will serve to illustrate how this method works. A Loran station consists of two transmitters, master and slave. Transmission of a signal from the master triggers the slave which also transmits. The observer must detect both signals and measure the time difference between them. His position is then fixed somewhere on a curve of constant time difference (a hyperbola, hence the name for this type of system). Observations of signals from two or more stations determine the position of the ship. A set of curves of constant time difference for one Loran station is shown in Figure 2.8.

In modern *satellite navigation* a position can be obtained to within a few hundred meters by observation of the passage of an artificial satellite. The orbit of the satellite must be determined by fixed tracking stations and this orbit must be known to the navigator. To determine his position he must measure two quantities; t_c and D_c, the time and distance of closest approach of the satellite to the ship. Knowledge of t_c and the satellite orbit places the observer somewhere on the line of intersection of a plane normal to the satellite velocity vector and the earth's surface. The position on this line is fixed by D_c. The necessary computations must be carried out on a computer; as with inertial navigation a large investment in equipment is required before the method is useful.

EXPLORATION OF THE SEA

Compared to the atmosphere and the land the oceans remain relatively unexplored. Investigation is more difficult and its importance not so obvious. Sea water is opaque to light except over short distances. Pressure increases at the rate of nearly a tenth of an atmosphere per meter of depth, placing a severe restriction on the depths to which divers can descend and complicating the design of underwater vehicles. A piece of equipment or a ship lost on the bottom of the deep ocean is lost forever as far as modern methods of salvage are concerned. Sound travels well through water, however, and sound pulses can be used to explore the larger features of the sea floor from surface ships. But more detailed studies, as with underwater photography, are so difficult and expensive as to be useful only in the examination of very restricted areas.

Scientific study of the deep ocean began with the HMS *Challenger* expedition of 1872–76, which made the first physical and biological survey of the Atlantic and Pacific oceans; research proceeded on a modest scale until the late 1950s when realization of the potential value of ocean resources (and of the strategic value of the seas) stimulated intensified exploration. Detailed mapping of the ocean floors has yet to be finished, while knowledge of oceanic circulation is much less complete than that of the atmosphere.

While it might be thought that a submarine would be the ideal vehicle for exploring the oceans, surface ships are more practical (and much less expensive) for much of the work of the oceanographer. A ship stopped on the surface can measure the temperature and salinity (salt content) of the sea water below as a function of depth, and the character of the sea floor. Temperature and salinity profiles are traditionally determined from water samples collected with Nansen bottles. A set of these bottles is spaced along a wire which is lowered over the side of the ship. When the wire is correctly extended a messenger weight is released from on deck; its arrival at

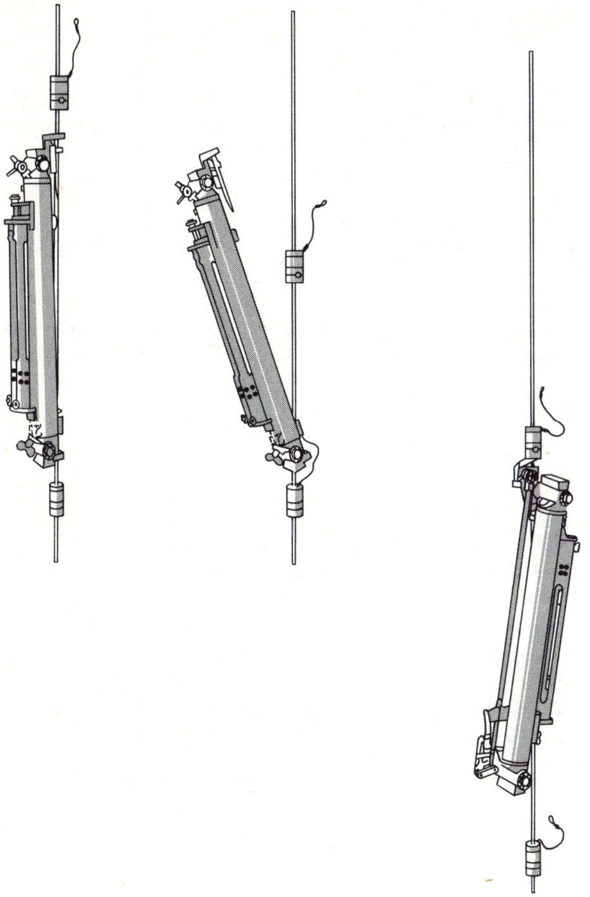

Before tripping Tripped After tripping

Figure 2.9 The operation of the Nansen bottle used to collect water samples below the surface. U.S. Hydrographic Office.

the first station on the wire causes the bottle to reverse, automatically closing its valves. Simultaneously, a second messenger is released to activate the next bottle below. The operation is illustrated in Figure 2.9. Temperature is read from the thermometers attached to each bottle; salinity is found by chemical analysis of a part of the water sample. When the temperature and salinity are known, the density of the sea water can be calculated. Density profiles obtained in this way are the basic data used to calculate oceanic circulation.

It would be desirable to measure the current flowing at each oceanographic station. It is easy to design an appropriate current sensor, but these instruments are almost impossible to use at sea. To measure current from a ship the vessel must be held in a fixed position. But it is impractical to moor a ship in the deep oceans and difficult to hold her on station with engines. A better solution is to use Swallow floats (named after their inventor). A *Swallow float* is a sphere about a meter in diameter whose weight has been carefully adjusted so that it will float at a predetermined depth below the sea surface. (This is possible because, with increasing pressure, the density of sea water increases with depth.) The heavier the float the deeper it will settle before

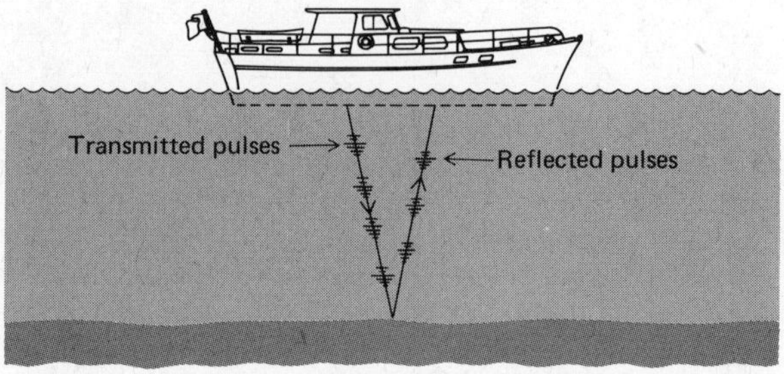

Figure 2.10 The fathometer measures water depth by measuring the time required by a sound pulse sent from the ship to be reflected from the bottom and return to the surface.

coming to an equilibrium depth. Floats are put over the side of the ship at a known position. As they are carried along by the ocean currents their motion is tracked either by locating the floats with sound pulses sent from the ship (as a submarine is tracked by sonar), or by having the floats emit sound signals from time to time. A sequence of observations of float positions establishes the direction and velocity of the current at the depth for which the float was set. The practical difficulty to be overcome is that of keeping track of the floats for a sufficiently long period of time. Adequate charting of the ocean currents is still far from complete.

While under way between stations an oceanographic ship can make a continuous record of water depth, thereby tracing out the profile of the sea floor. This is accomplished with an echo sounder or *fathometer.* The principle of its operation is illustrated in Figure 2.10. A sound generator mounted in the ship's hull sends out sound pulses at regular intervals. These are reflected by the sea floor and return to the ship as echoes. The deeper the water, the more time will elapse before an echo is returned. The time interval, t, between transmission and echo is measured and used to find the water depth, d, by the relation

$$d = \frac{vt}{2} \tag{2.2}$$

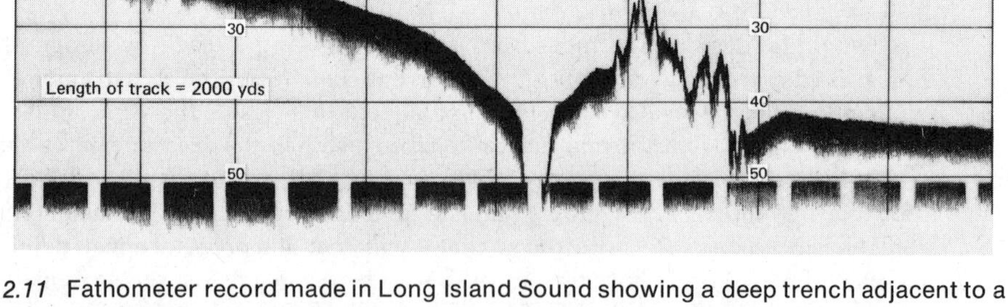

Figure 2.11 Fathometer record made in Long Island Sound showing a deep trench adjacent to a shoal area. The track starts near shore (left) and proceeds seaward (to the right).

where v is the velocity of sound in sea water. A fathometer records the depth so determined on a continuously moving chart. Figure 2.11 is an example. In addition to reflection from the bottom, sound in the sea may also be reflected by schools of fish or even layers of water at different temperatures. False readings—apparent shoals—are not infrequently recorded by fathometers, but are often useful to fishermen.

The structure of the sea bottom can be photographed by underwater cameras

Figure 2.12 A sediment corer to be put overside. With permission of Lamont Geological Observatory.

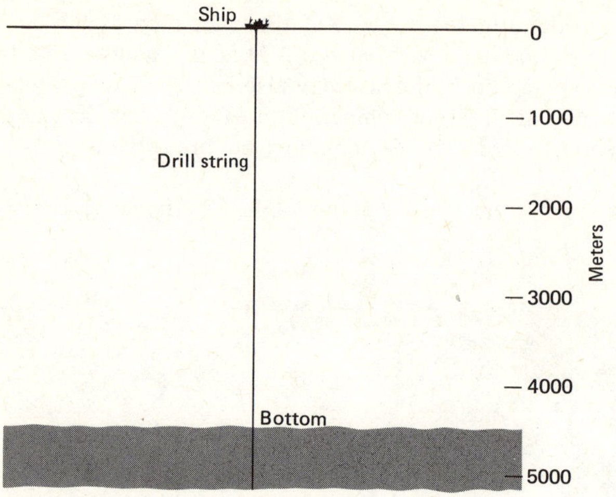

Figure 2.13 The *Glomar Challenger* with its drill string shown to scale.

lowered on cables, or, better, observed directly from a deep submergence vehicle such as the bathyscaphe *Trieste*. The *Trieste* has the capability of reaching the bottom of the deepest ocean. The sea floor is found to be covered in most places by a layer of sediment, fine particles of mud, sand or animal remains that have settled to the bottom. Sediment samples are obtained by surface ships by the use of a corer. A corer is a hollow tube weighted so that it will penetrate the sea floor (Fig. 2.12). Fitting the core tube with a piston that may be drawn up as the tube penetrates the bottom aids in getting the sediment to fill the tube. After the core tube is filled the corer is hauled to the surface and the core of the sediment removed for study. Cores with a length up to about 10 meters can be obtained with this type of gravity corer. Deeper penetration requires drilling. Underwater drilling for oil has long been practiced in relatively shallow waters. The techniques have recently been refined to allow drilling in the deep oceans. The deep-sea drilling ship *Glomar Challenger* is one of the outstanding technical developments of the decade and cores she has obtained are of critical importance in understanding the formation of the ocean basins (as explained in Chap. 8). Often this ship's drills can penetrate entirely through the sea floor sediments and bite into the bedrock underneath. In water as much as 5000 meters deep, cores up to 1000 meters long can be recovered. Figure 2.13 gives an idea of how long the drill string is, compared to the size of the ship.

The explorations of the last decade have resulted in an enormously increased understanding of the oceans. The next section describes the form of the land beneath the sea; a discussion of oceanic circulation follows.

OCEAN BASINS

Shown in Figure 2.14 is the percentage of the earth's surface in each 1000 meter interval above or below sea level. There are two well-defined maxima, one for elevations between 0 and 1000 meters, the other for depths between 4000 and 5000 meters. This means that the earth's surface can be divided into the oceanic areas (averaging 3800 meters deep) and the continental areas (averaging 840 meters above

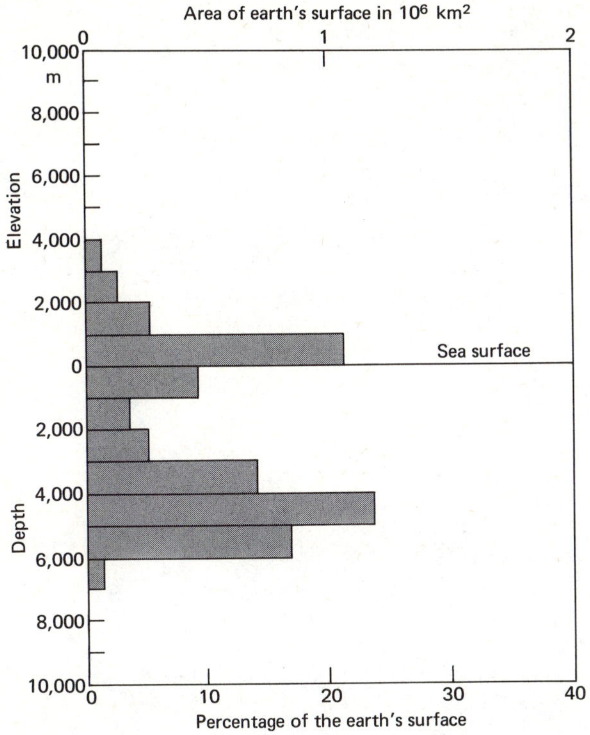

Figure 2.14 Each bar shows the amount of the solid earth surface in the corresponding 1000 meter range of elevation. Note that there are two well separated peaks in the distribution; these correspond to the continental and oceanic areas of the world. After F. D. Stacey, *Physics of the Earth,* copyright 1969 by John Wiley, based on the data of E. Kossinna.

sea level), with a fairly sharp division between the two. The relatively steep boundary between continental and oceanic areas is the *continental slope.* It can be clearly seen in the relief map of Figure 2.15. The seaward margins of the continents which are covered by the sea are the *continental shelves,* whose geologic structure is like that of the continents themselves. They are a part of the sea floor on which resources such as natural gas and petroleum can be expected to occur. Offshore, across a continental shelf, the water deepens gradually until a depth of about 200 meters is reached, then the bottom drops away suddenly to the approximately 5000 meters depth characteristic of the deep ocean. Under the sediments of the ocean basins is a base of volcanic rock in which there is every reason to believe few valuable mineral resources will ever be found. A map of the world constructed to distinguish continent from ocean should be drawn around, say, the 1000 meter contour, midway down the continental slope rather than around the shoreline. Figure 2.16 shows the world map redrawn this way. Major continental shelves are the Grand Banks, the land under the North Sea, the Sea of Arafura, and the land off the Patagonian coast of South America.

Until not very long ago it was generally thought that the floor of the deep ocean was a flat, featureless plane. It is not. Indeed, its topography is more rugged than that of the continents, being marked by high undersea mountains and deep valleys called trenches. The deepest ocean trenches are substantially deeper than the highest mountains are high. The mountain belts of the oceans are much more nearly continuous than those of the continents. They form a chain running generally through the central region of each ocean basin. This is known as the mid-ocean ridge system. A

Figure 2.15 A perspective drawing with vertical exaggerations of the North Atlantic ocean basin. Courtesy of the National Geographic Society.

few peaks are high enough to break through the sea surface, as do St. Peter and St. Paul Rocks and Ascension Island, a mountain peak rising 6 km above the sea floor. The center of the mid-ocean ridge is generally marked by a sharp valley between high neighboring mountains, as may be seen in Figure 2.15. This figure also shows that the ridge axis is frequently offset laterally, giving the mountain chain a discontinuous appearance. Associated with each offset is a *fracture zone* which extends perpendicular to the ridge axis. The world-wide ridge and fracture zone system is charted in Figure 2.16.

Plunging below the 5 km average depth of the sea floor are ocean trenches, narrow, deep valleys where depths exceed 10 km. An example is the Puerto Rico trench in Figure 2.15. As Figure 2.16 shows, ocean trenches are found near the ocean margins and are not so continuous a network as the ridges.

Away from the central mid-ocean ridge the sea bottom tends toward a flat, featureless floor called the *abyssal plain*. In some places this plain is pierced by isolated mountain peaks which are usually extinct volcanoes. The Cape Verde Islands, shown in Figure 2.15, are examples. The abyssal plains are mantled by a covering of sediment up to several hundred meters thick. This sediment is composed partly of fine mineral grains washed into the sea by rivers but mostly of skeletal remains of tiny sea creatures. Microscopic shells of dead animals are continuously falling to the sea floor where they accumulate as sediment. When a section of this sediment is examined under the microscope the animal species responsible for the remains can be identified. Such observations can be used to establish the time at which any particular layer of sediment was formed. Over periods of thousands of years significant changes in animal structures occur by evolution through natural selection; the time when any particular species was alive is generally known to paleontologists. From the animal species present, then, the time of formation of a sediment layer can be determined. These methods have been used to great advantage in the study of deep sea cores obtained by the *Glomar Challenger*. Particularly interesting is the age of the bottom layer of sediment (the layer resting on the underlying igneous rock floor). It might be anticipated that this would be the same throughout the oceans, corresponding to the time when animal life first appeared in the seas. This is not the case. First, the age of the bottom layer of sediment regularly increases outwards from the mid-ocean ridges, and second, the greatest age ever found is about 100 million years. There has been life in the sea for at least ten times that long. Shown in Figure 2.17a are the locations of a series of cores collected in the South Atlantic Ocean relative to the mid-ocean ridge, and in Figure 2.17b, the age of the bottommost layer of sediment from each core. The regular increase of age with distance outwards from the ridge is evident. These are crucial data for understanding the evolution of the oceans; they show that the ocean basins are relatively young features on the face of the earth (whose age is 4500 million years, see K. K. Turekian's *The Chemistry of the Earth*). We return to this topic in Chapter 8.

SEA LEVEL

We have seen that the earth's surface is divided into continents and ocean basins with a fairly sharp boundary between the two. It is remarkable that the amount of water available on earth is just about right to fill the ocean basins. Actually, there is a slight excess so that the seaward margins of the continents are flooded (forming the conti-

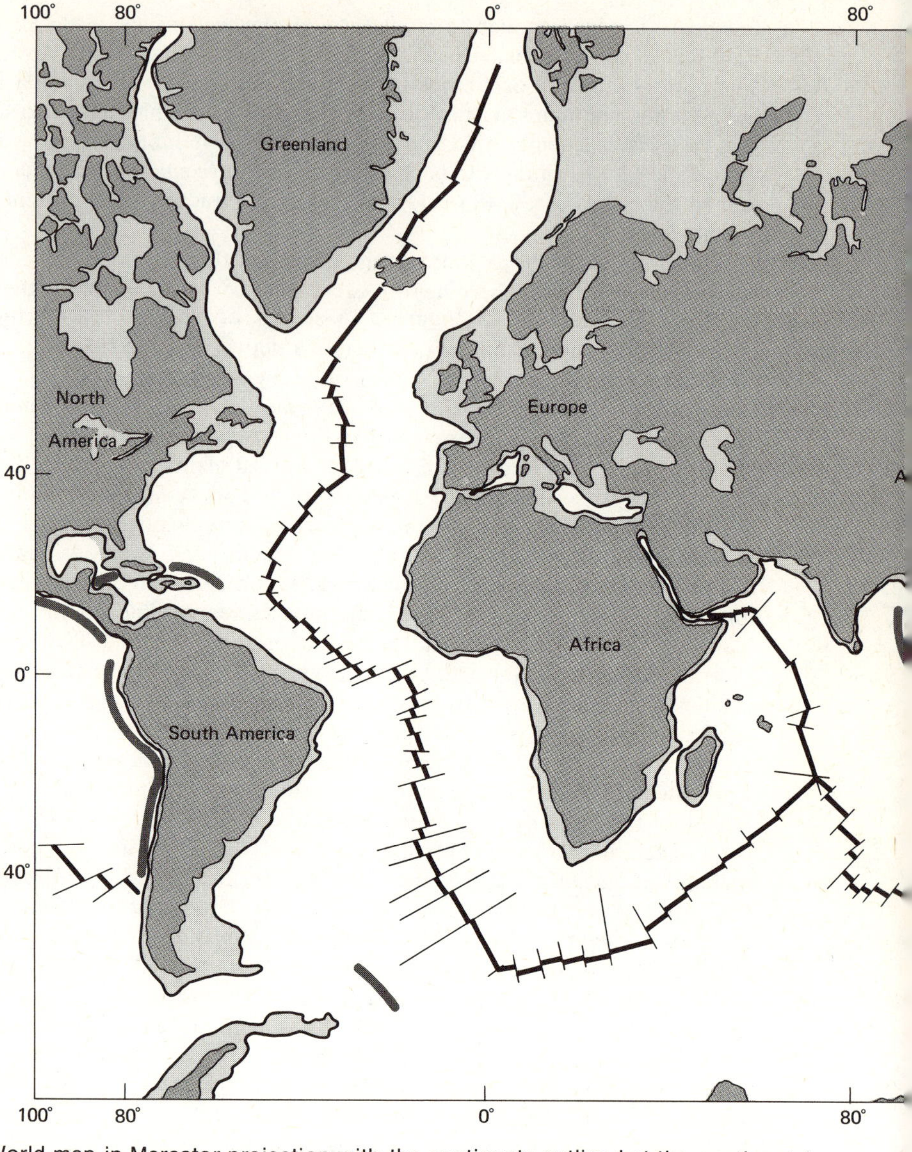

Figure 2.16 World map in Mercator projection with the continents outlined at the continental slope (the 1000 meter contour). This shows the true shape of the continental areas and the extent to which the continent margins are now flooded by the sea. Also shown is the

nental shelves). A substantial amount of water is now locked up in the ice caps of Greenland and Antarctica. Were this ice to melt, sea level would be raised by about 100 meters and the continental margins would be more deeply submerged. Sea level

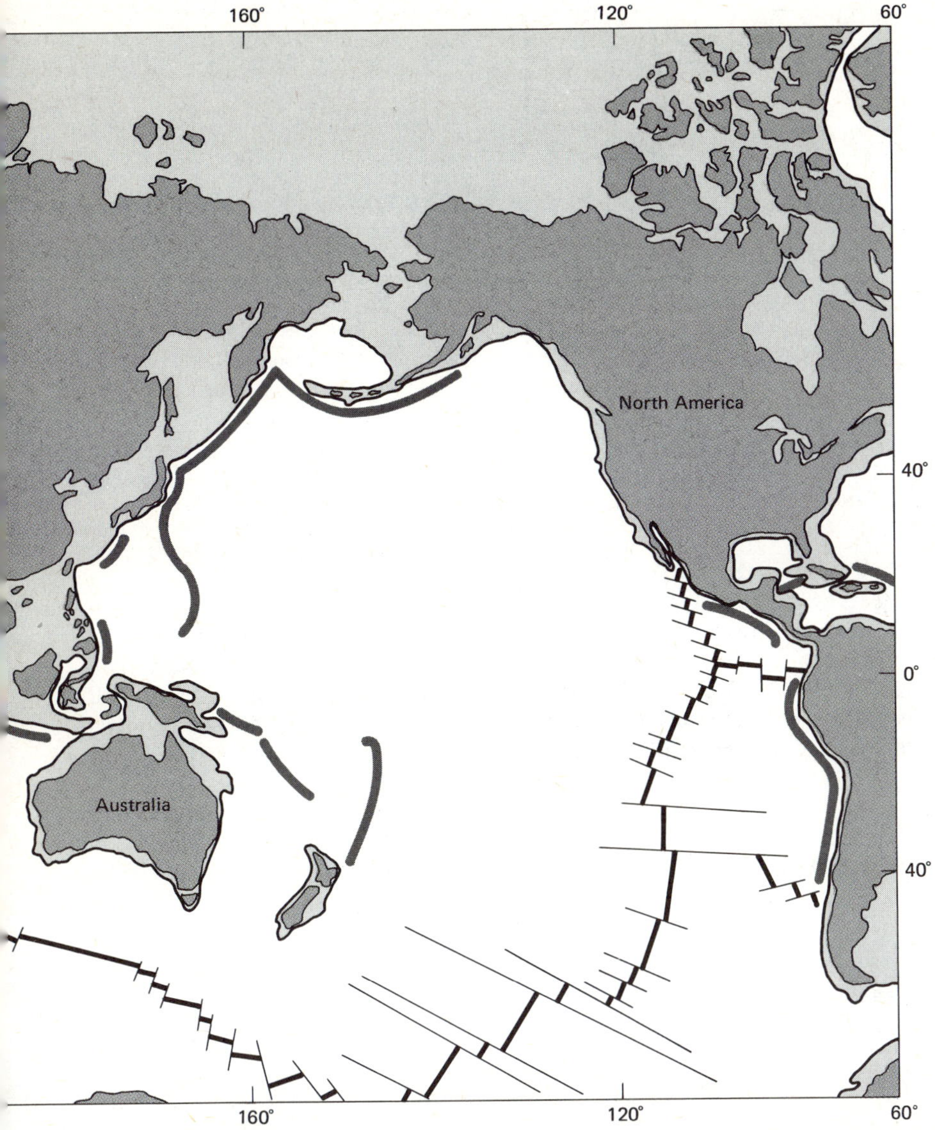

ocean ridge system (heavy line) and its associated fracture zones (light lines). Ocean trenches are shown in dark gray. Based on BA chart 3934 with the sanction of the Controller, HM Stationery Office and of the Hydrographer of the Navy.

is, in fact, not constant; it appears at the present time to be rising. Mean sea level can be found from the tide gage records kept at many ports throughout the world. At ports on the east coast of the United States, sea level is rising at the rate of about 3

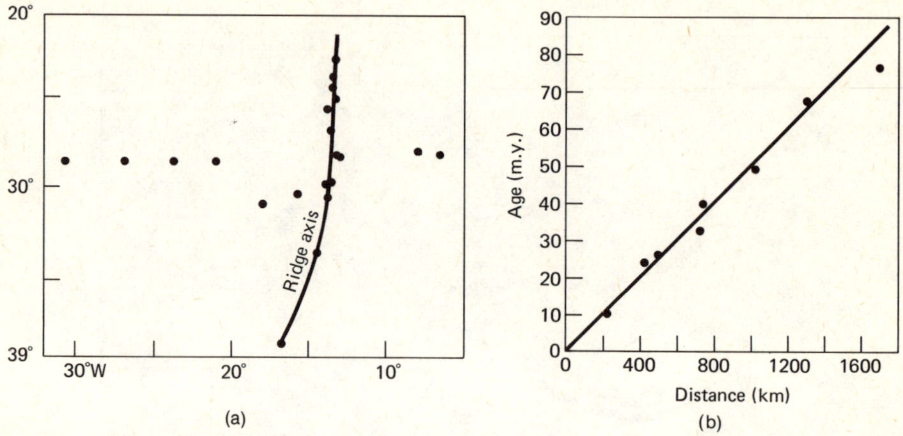

Figure 2.17 (a) The location of a set of drill cores relative to the mid-ocean ridge in the South Atlantic, and (b) the age of the bottommost layer of sediment from each core plotted against distance from the ridge axis. After Maxwell *et al.*, *Science* **168,** 1047 (1970).

mm per year. (A small part of this rise is due to sinking of the coast but most is apparently due to an increase in the amount of water in the oceans.) The rise in sea level has been going on for at least 10,000 years. This is illustrated by the data in Figure 2.18 which shows the rate of infilling of a salt marsh on the Connecticut shore. Salt marshes are useful in measuring sea level because sediment accumulates to mean high tide level and no higher. The upward growth of the marsh records the rise of sea level.

The low sea level which existed 10,000 to 15,000 years ago was a consequence of the last ice age. During the ice age the polar ice caps became much larger, the ice in eastern North America extending down to about the latitude of New York. The

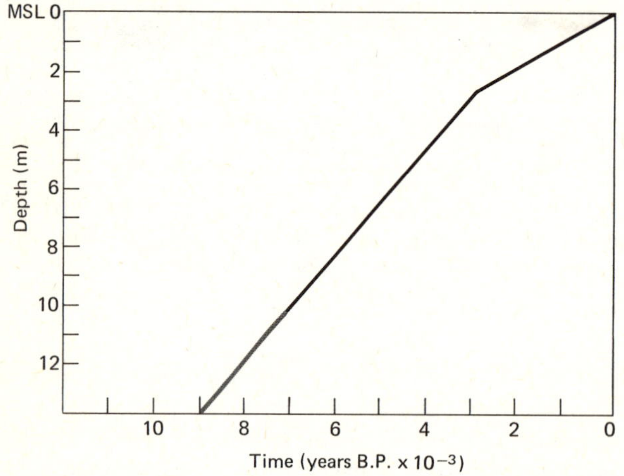

Figure 2.18 Sea level on the Connecticut coast over the past 8000 years as determined from the study of the rate of infilling of a salt marsh. After Bloom and Stuiver, *Science* **139,** 332 (1963).

water required to make this ice came from the sea and sea level was therefore lowered. At the height of the ice age, most of the continental shelves were above sea level. When the earth started warming after the last ice age, rivers carrying water from the melting glaciers cut deep valleys in the continental margins. These were subsequently flooded by rising sea level and are today submarine canyons. An example is the Hudson canyon, a seaward extension of the valley of the Hudson River. Thus, the topography of the continental shelves was initially shaped as dry land features. George's Bank, the important fishing bank off Cape Cod, was formed, for example, as a deposit of rubble dropped by the glacial ice.

During the low stand of sea level in the last ice age, many sea mounts projected above the ocean surface. As sea level rose with the melting of the ice caps these sea

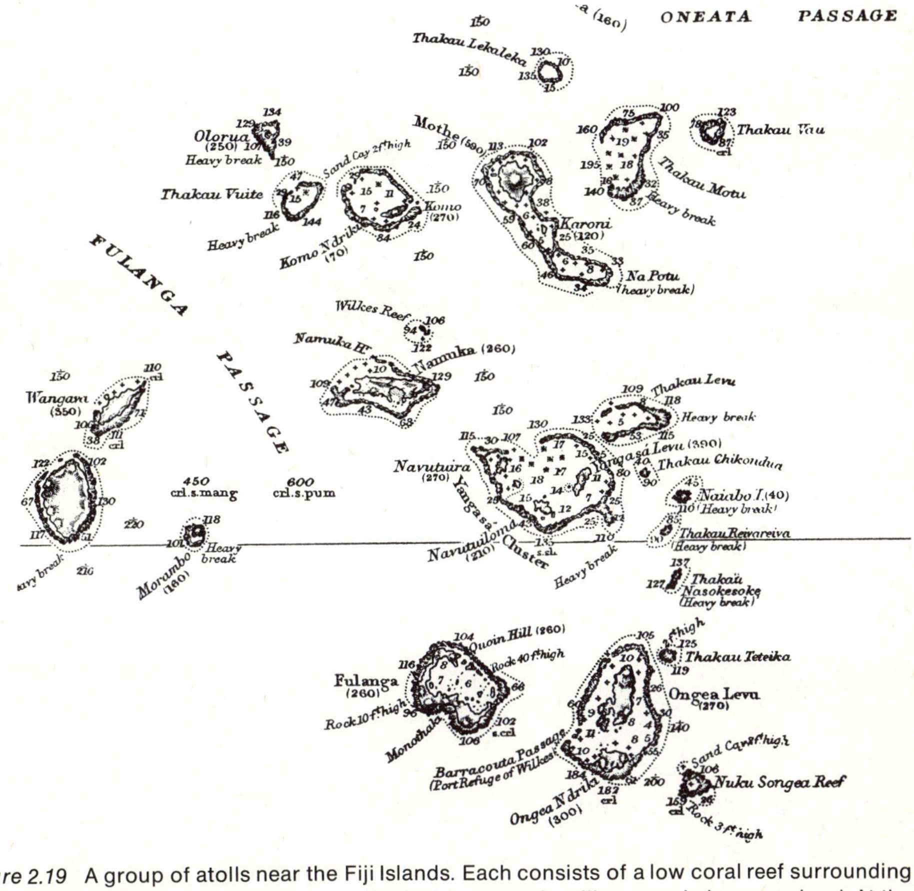

Figure 2.19 A group of atolls near the Fiji Islands. Each consists of a low coral reef surrounding a shallow lagoon. At Mothe the peak of a sea mount is still exposed above sea level. At the other islands the mountain peaks have been covered by the sea and only the coral reef remains at the surface. Reproduced from BA chart number 2691 with the sanction of the Controller, HM Stationery Office and the Hydrographer of the Navy.

mounts were submerged. But in tropical climates coral can grow rapidly in shallow water. When sea level is rising, coral grows upward fast enough to remain near the surface. The more deeply submerged coral eventually dies but the living coral continues to grow on it. The result of this process is the coral atoll familiar in the South Pacific. It typically consists of a low coral reef surrounding a shallow lagoon. Outside the reef the bottom drops away very rapidly to the depth of the main sea floor. These features can be seen in the chart of a group of atolls near the Fiji Islands (Fig. 2.19).

The practical significance of sea level changes today is the change in the seacoast which results. Many of the major metropolitan areas of the United States are only a few meters above sea level. Over the next few hundred years the sea will (if the present rise continues) encroach on these areas to a significant extent. Where the coast is flat and has a low slope much land will be lost to the sea. If the average slope of a shore is 1/100, the present rate of sea level rise means the submergence of 0.3 meter of shoreline per year. In a decade a shorefront property owner can expect to lose a significant amount of land.

SEA WATER

Sea water is salt water because minerals dissolved by rain water on the continents are carried to the oceans by rivers. It is also cold water, all but a surface layer being at temperatures only a few degrees above freezing. Figure 2.20 shows the temperature distribution found in the tropics and in the temperate zones. Radiant heat received at the sea surface is used partly in evaporating and partly in heating the water. Wave action keeps the surface layer of water well mixed so that there is a warmed surface layer (all year in the tropics, summer only in the temperate zone) extending to a depth of about 200 meters. The rapid decrease in temperature at the base of the warmed layer is called the *thermocline*. Below it the water is uniformly cold and remains so even to the bottom of the deepest trenches.

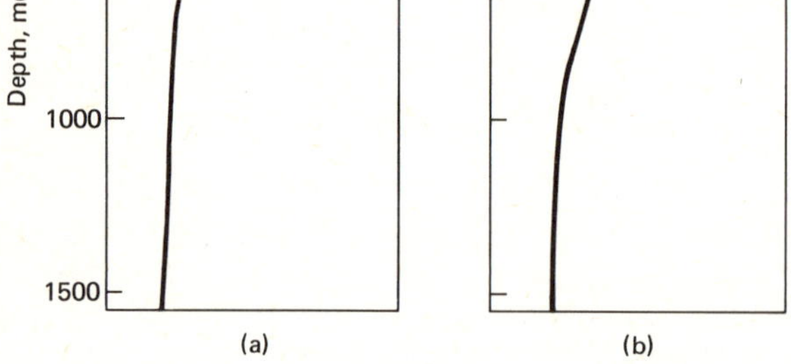

Figure 2.20 The temperature profile of the open ocean (a) at low latitudes and (b) at mid-latitudes.

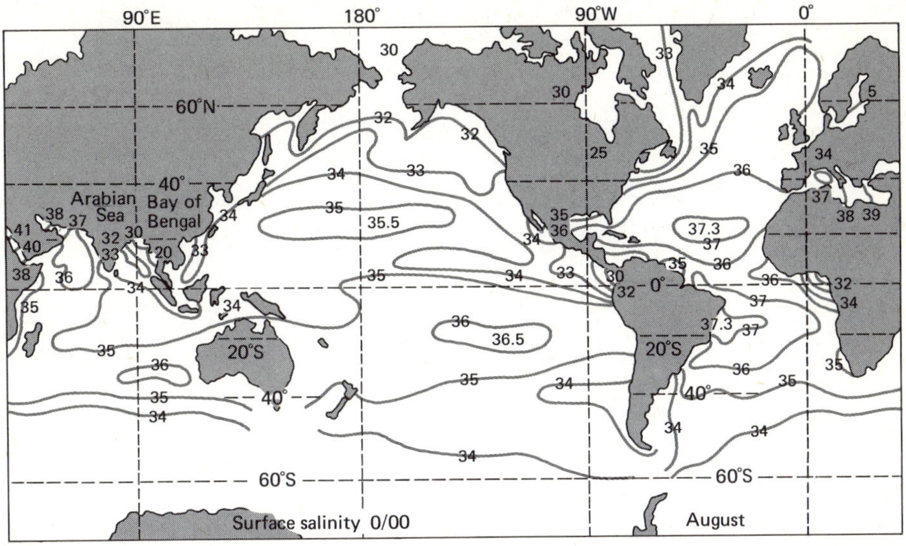

Figure 2.21 Surface salinity of the oceans during August. From data of Pickard, *Descriptive Physical Oceanography.*

The salt content of sea water is called its *salinity*. Throughout the deep oceans it is about 3.5 grams of salt per 100 grams of water or 35‰ (parts per thousand). The distribution of surface salinity in the oceans is shown in Figure 2.21. Low salinity is found where there is heavy discharge of fresh water into the sea from rivers; high salinity, where there is much evaporation and little fresh water supply, as in the Red Sea. The vertical distribution of salinity is influenced strongly by evaporation; the surface layers are of high salinity in the tropics (Fig. 2.22). Variations in salinity are important to marine life as many species can tolerate only a narrow range of salt concentration.

Since salinity and temperature are the principal factors controlling density, they affect the circulation of sea water. Water movement resulting from changes in temperature and salinity is called *thermohaline* circulation. It does not occur on the same scale as the thermal circulation of the atmosphere. This is because solar radiation heats both the ocean and the atmosphere only very near the earth's surface. Air heated at ground level acquires a lower density than the air above, so that the bottom layer of the atmosphere becomes unstable. This initiates convection. In the oceans, the warm, low density water is already on top; further heating cannot cause it to rise. In the tropics the salinity of surface water is increased by evaporation, but as the water is also heated, extensive circulation does not develop from this cause. Circulation is affected only when the surface water becomes denser than that below. For example, an important source of thermohaline circulation is the freezing of the sea surface in polar regions. Sea ice contains little salt. As freezing proceeds, the water ahead of the advancing ice is enriched in salt. Its density increases and it drops down to lower depths, becoming "bottom water" which slowly drifts along the sea floor from the poles to the equatorial zone.

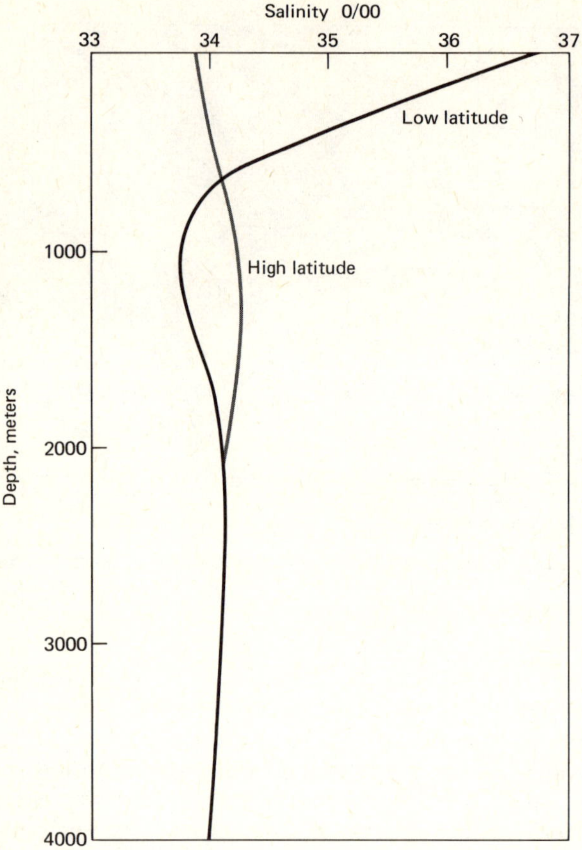

Figure 2.22 Salinity profile of the open ocean at two latitudes.

WIND-STRESS CIRCULATION

Wind blowing over the sea exerts a drag stress on the surface which sets the surface layers in motion. The resultant flow, modified by the Coriolis force (Chap. 1), constitutes most of the large-scale circulation in the oceans. A simple example of wind-driven circulation is the upwelling of cold water off the west coast of South America. The high Andes Mountains are close to the coast and deflect onshore winds northward, parallel to the coast and towards the equator. The wind drag force at the water surface acts towards the north but, because the Coriolis force in the southern hemisphere is to the left, the surface water moves to the west, away from the shore. Its place is taken by the water which moves upward from depths of 50 to 300 meters. This water is relatively cold (too cold for swimming) but it is rich in the nutrients, phosphates, nitrates, and silicates, required by marine life. For this reason, the coastal waters, particularly off Peru, are extraordinarily productive of fish; they support one of the world's largest fisheries. The abundant fish, in turn, support a large population of sea birds. The guano produced by these birds has long been an impor-

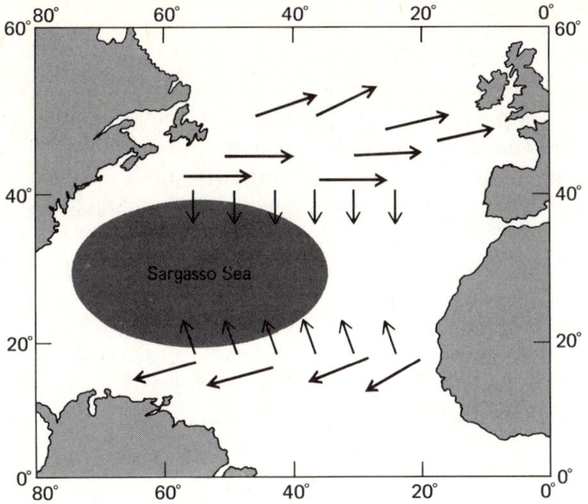

Figure 2.23 Prevailing wind pattern (black arrows) over the North Atlantic. The surface drag of these winds and the Coriolis force produce a flow of water (gray arrows) into the Sargasso Sea. From *Scientific American* (Sept. 1969).

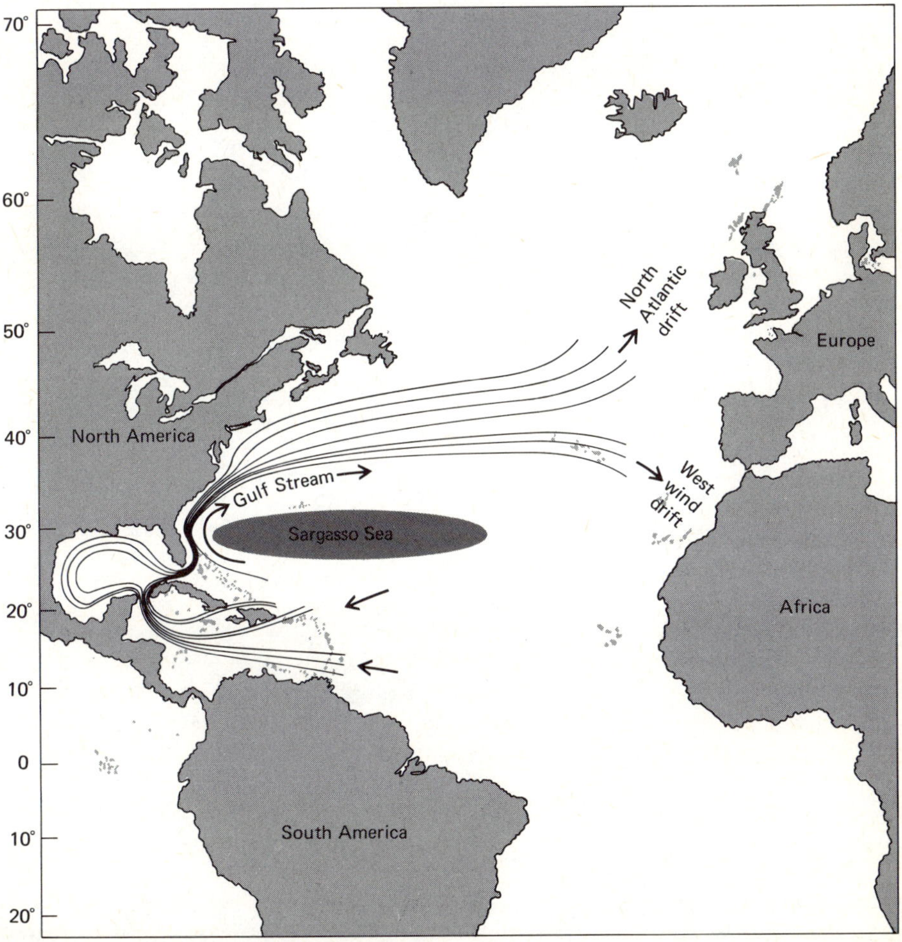

Figure 2.24 The track of the Gulf Stream.

tant source of phosphate-rich fertilizer. Thus the minerals washed into the sea by erosion are returned to the land for further use. (Today, unfortunately, overfishing poses a threat to the fishery, the bird population, and the production of guano.)

The general circulation of surface waters in the North Atlantic is a consequence of the northeast trades blowing across the southern portion and the prevailing westerlies over the northern portion (Fig. 2.23). The drag of these wind systems and the Coriolis force direct the surface waters of the North Atlantic into the central area known as the Sargasso Sea, from which clearly there must be a return flow of water into the northern portion of the ocean. This return flow is strongly concentrated in the western part of the North Atlantic. It is the Gulf Stream. A chart of its track is given in Figure 2.24. Part of the northward flow is initiated through the Caribbean and the Gulf of Mexico and enters the Atlantic as the Florida current. This is joined by the Antilles current to form the Gulf Stream. The Stream consists of a number of

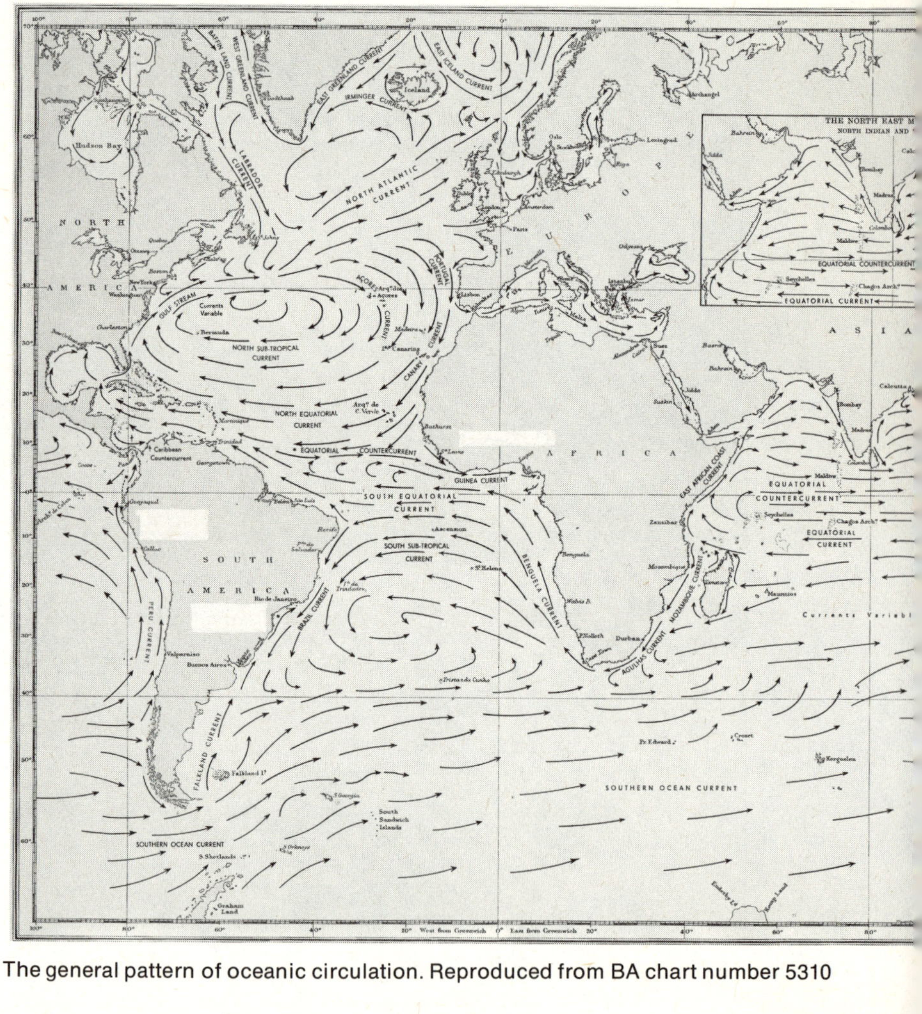

Figure 2.25 The general pattern of oceanic circulation. Reproduced from BA chart number 5310

filaments ranging to a depth of about 400 meters and moving with velocities up to 5 knots (fast enough to have an important effect on the navigation of ships). The boundary between the warm, blue water of the Gulf Stream and the cold, green water to the north is sharp and is often easy to see from the air. The position of the stream meanders about and its exact location at any particular time cannot be predicted. A vessel making passage from northwest to southeast should take frequent water temperature measurements; allowance for a transverse current of several knots must be made when a sharp temperature rise is detected. In the North Atlantic the boundary between the warm and cold water is marked by frequent and dense fog (for which the Grand Banks are well known). The warm water transported to the north by the Gulf Stream has an important climatic influence on Europe, and particularly the British Isles, giving these much warmer average temperatures than North America at the corresponding latitude (Labrador).

with the sanction of the Controller, HM Stationery Office and the Hydrographer of the Navy.

OCEANS AND THE ATMOSPHERE

There is a close connection between the large-scale circulation of the oceans and the atmosphere. The drag on the sea surface by the prevailing wind pattern is responsible for large-scale circulation of surface waters in each ocean basin (Fig. 2.25; compare Fig. 1.6); oceanic circulation is driven by the winds. But in return, much of the energy that drives the winds is transmitted to the atmosphere through the agency of the oceans. Solar radiation on the sea surface evaporates water into the air. Condensation of this water releases heat in the atmosphere, providing energy to drive the winds. Ocean and atmosphere are coupled systems. Changes in the circulation of one will change the circulation of the other. Thus the oceans play a large role in regulating climate.

APPENDIX: THE MARINE SEXTANT

Successful celestial navigation requires precise measurement of altitudes of celestial bodies. A remarkable degree of precision is attained in altitude measurements with an instrument as simple as the marine sextant. It consists of a rigid frame to which is fitted the fixed mirror M_1, and the mirror M_2 which can be rotated about the pivot at O (Fig. 2.26). The mirror M_1 is silvered over half of its width only. On looking through the telescope, T, then, the observer can look past the silvered part of M_1, at the horizon, H, and simultaneously see the image of the celestial body reflected by M_2. With some practice the two images can be made to fuse so that the view is like that shown. To measure the angle α, M_2 is rotated until the celestial body appears through the telescope to be just tangent to the horizon. The amount of rotation of M_2 is indicated by the pointer on the scale, S. (The instrument is known as a sextant because the arc of S is 60°; angles of elevation up to 120° can be observed because $\alpha = 2\theta$.) With the aid of a micrometer screw on the end of the pointer arm, α can be measured to $0'.1$ (a tenth of a minute of arc).

Under ideal atmospheric conditions, then, it is possible with this simple, hand-held instrument, to determine distance from the equator to within 200 meters. However, the index of refraction of air varies with density and density varies with elevation above the earth's surface. Traveling to the observer, O, in Figure 2.27 at a height,

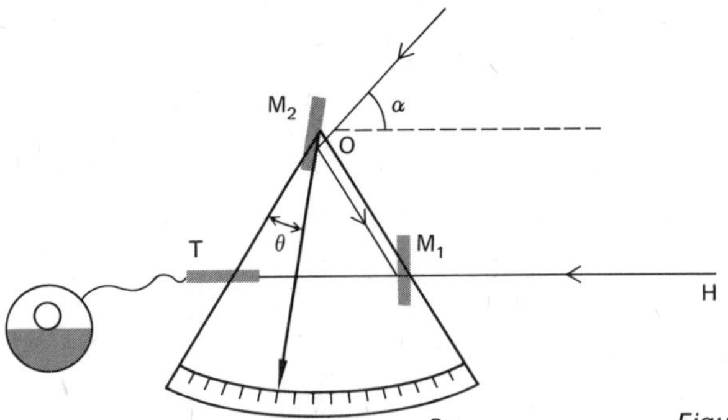

Figure 2.26 The marine sextant.

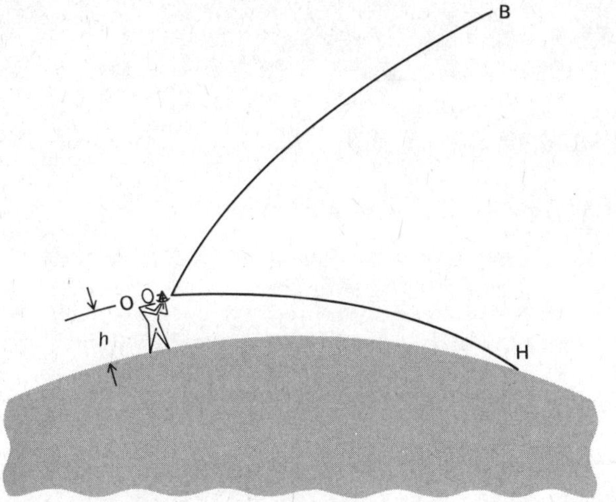

Figure 2.27 Light rays reaching the observer from the celestial body, B, and the horizon, H, are refracted by the atmosphere and so follow curved paths.

h, above the surface (usually the height of the navigating bridge of a ship above sea level), rays from the horizon and from the observed body are bent because of the nonuniform refraction. Corrections for this bending are calculated on the basis of a standard atmosphere, but of course when the atmosphere is nonstandard, the corrections are invalid. The greatest error arises when the air is colder than the water. Then air heated at the ocean surface is unstable and flows upwards in convection currents. The resulting inhomogeneity of the atmosphere obscures the horizon, which appears to "boil" and causes an erroneous measurement of α.

Further Readings

DEACON, G. E. R., ed., *Seas, Maps, and Men.* London: Crescent Books, 1962.

HEEZEN, B. C., and HOLLISTER, C. D., *The Face of the Deep.* New York: Oxford, 1971.

PICKARD, G. L., *Descriptive Physical Oceanography.* London: Pergamon Press, 1966.

RITCHIE, G. S., *The Admiralty Chart.* New York: Elsevier, 1967.

SVERDRUP, H. W., JOHNSON, M. W., and FLEMING, R. H., *The Oceans, Their Physics, Chemistry and General Biology.* Englewood Cliffs, N.J.: Prentice-Hall, 1946.

THOMSON, SIR C. W., *The Atlantic: The Voyage of H.M.S. Challenger.* New York: Harper, 1878.

TUREKIAN, K. K., *Oceans.* Englewood Cliffs, N.J.: Prentice-Hall, 1968.

TUREKIAN, K. K., *Chemistry of the Earth.* New York: Holt, Rinehart and Winston, 1972.

VON ARX, W. S., *An Introduction to Physical Oceanography.* Reading, Mass.: Addison-Wesley, 1962.

3 Tides: Physical Analysis of a Natural Phenomenon

"But the tide (which stays for no man) calling them away, that were loathe to depart, . . . [they] took their leaves of one another." So did Nathaniel Morton record the departure of the Pilgrims for the New World. The state of the tide has long been a matter of comment, and of practical concern in the operation of ports and harbors and the navigation of ships in coastal waters. It has also been the subject of detailed scientific study; its main features are now well understood and can be explained in terms of basic, simple physical principles.

The tide is the regular rise and fall of sea level. In most places there are two high and two low tides each day, the ebb and the flood of the tide each requiring a little more than six hours. The height of the tide (also called its range) is as great as 20 meters at the head of the Bay of Fundy, from 3 to 10 meters at many important seaports, but nearly zero in the Mediterranean Sea. Tidal currents accompany the rise and fall of water level. In some places, as in the Pentland Firth north of John O'Groats, these currents run as fast as 10 knots (about 11 mph), fast enough to be a serious problem to all but the most powerful ships. Tidal currents control the bottom structure of estuaries and bays and through this their animal population. They are often accompanied by *tide rips,* strong turbulence which can be an important hazard in navigation.

The prediction of tides has always been of interest. Advance knowledge of the times of high and low water is essential to the operation of a port such as Boston or Liverpool; it affects sailing times and positions for docking. Before Lord Kelvin achieved a scientific understanding of the tides in 1867, predictions were made by guesswork methods which were often closely held as family secrets over many generations. Now the tides are predicted by computers operated by agencies such as the U.S. National Ocean Survey and Admiralty Hydrographic Office. Scientific interest today is turned toward the effects of tidal friction and the use of the tides to generate electric power.

THE TIDE-PRODUCING FORCE

The gravitational attractions of the sun and moon, and the rotational forces due to their relative motion, act on the water of the oceans to produce the tides. The same forces act on the air and the land, producing atmospheric and earth tides, but only the oceanic tides can be observed without special instruments.

Gravitational forces are responsible for the attraction of one mass for another. All masses attract one another; Isaac Newton found that the gravitational force between two bodies decreases in proportion to the square of the distance between them and increases in proportion to the product of the masses. Thus, Newton's law of gravitation states that a particle of mass m_1 attracts a particle of mass of m_2 a distance r away with a force

$$F = G \frac{m_1 m_2}{r^2} \tag{3.1}$$

The constant of proportionality, G, is called the gravitational constant. It is a small quantity; the gravitational force between two 1 kg masses one meter apart, for example, is so small that extraordinarily delicate instruments are required to detect it. If m_1 is the mass, and r the radius of the earth, F is the weight of a mass m_2 on the earth. This same mass would weigh less on the moon because of the moon's smaller gravitational attraction.

If the earth and moon were very far apart compared to their sizes, the gravitational force exerted by the moon on an object on the earth's surface would be nearly the same at all places on the earth. Actually, the ratio of the average earth–moon distance, R, to the radius of the earth, r, is $R/r = 60.26$. Thus, in Figure 3.1 point U on the earth is significantly closer to the moon than is point V and the gravitational attraction of the moon for unit masses at U and V will be appreciably different. An object at U will weigh slightly less than the same object at V.

The earth and moon together are a dynamic system revolving about point A in the figure (which is their common center of gravity). The gravitational force of attraction between them is balanced by the centrifugal force due to their revolution. The earth also rotates about its own axis, but this rotation makes no contribution to the forces which raise the tides. The two motions of the earth—its revolution about point A and its rotation about its axis—may be considered separately. The tides result from the gravitational attraction between the earth and moon and their revolution about point A only. Hence, for the tidal calculation the earth may be thought of as not rotating at all. As the moon and the earth revolve about point A, the center of the earth, C, follows the circular path shown by an arrow. If the earth were a rigid body every point on earth would be rigidly attached to every other point; hence, every point on

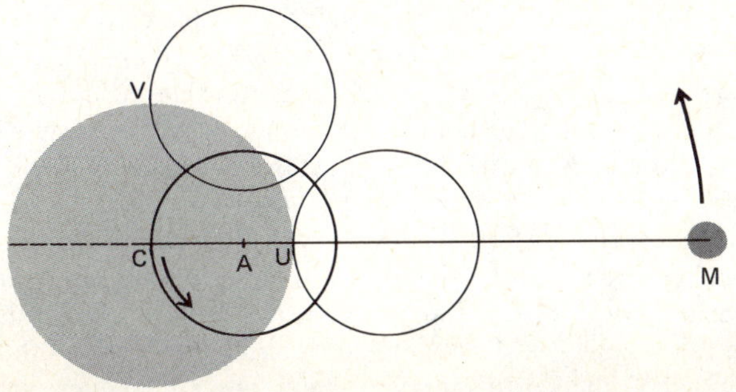

Figure 3.1 The earth, with center at C, and the moon (M) revolve about their common center of gravity, A, so that C traces out the heavier black circle. Points U and V, and every other point in the (non-rotating) earth, follow a circular path of the same radius.

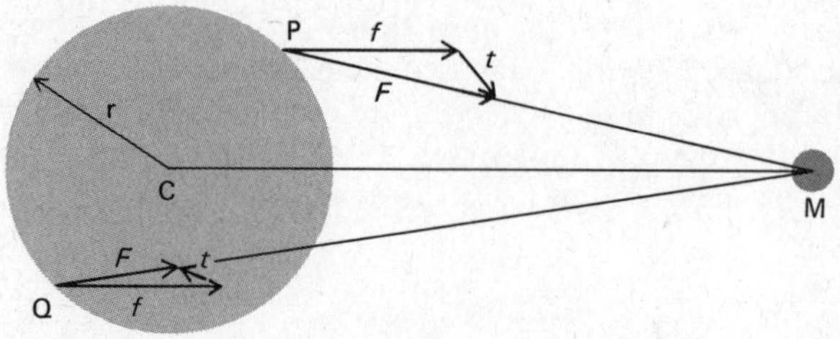

Figure 3.2 The tide-producing force, *t*, at points P and Q on the surface of the earth is the difference between the gravitational force of the moon, *F*, and the force, *f*, required to maintain motion of a particle in a circular orbit. The difference is found by vector subtraction.

or in a nonrotating earth would follow a circular path of the same radius at the same velocity as that followed by C as it revolves around A. For example, the circular paths for points U and V are shown in Figure 3.1.

In order to keep a particle on a circular path, there must be a restraining force acting inwards along the radius of the path which just balances the centrifugal force. For a particle moving around one of the circles of Figure 3.1 this force will be designated as *f*. For the particle at C, *f* is provided exactly by the gravitational attraction of the moon. At any other point in the earth the gravitational force provided by the moon will not be quite right in direction or magnitude to provide *f* exactly. This is illustrated for points P and Q in Figure 3.2. The magnitude and direction of *f* must be the same at P and Q because both P and Q are moving on circular paths of the same radius at the same speed. But, at neither P nor Q is the gravitational force due to the moon, *F*, in the required direction (the direction of *f*). Furthermore, at P the lunar gravitational force *F* is too big, because P is relatively closer to the moon, whereas at Q, which is further away, *F* is too small. The difference between the vector *f* and the vector *F* is the additional force that keeps P or Q moving on their circular paths. In the rigid parts of the earth this additional force is supplied by the strength of the earth's material. Points P and Q move exactly on circular paths of the same radius and no relative motion of P and Q occurs.

Now suppose that P and Q are points in the ocean. Then there is no way of supplying the additional force to keep them on circular paths because the water has no rigidity. Each of these points acts as if it were subjected to the force *t*, the difference between forces *F* and *f*. The magnitude and direction of *t* is found by vector subtraction (see Appendix I) according to the construction shown in Figure 3.2. The forces *t* are the tide-producing forces. They arise because of an imperfect match of the centrifugal force *f* and the moon's gravitational force *F*.

The tide-producing force, *t*, will be different at every point on the surface of the earth. Its variation around a great circle passing through the earth's poles is shown in Figure 3.3. At the poles *t* acts vertically downwards, at the place where the circle intercepts the equator, vertically upwards. At these places the tide-producing force causes only a change in the weight of a particle. This is because the force *t* acts in the same, or the opposite, direction as the earth's gravity. At the poles it acts in the same direction as the earth's gravity and the particle is heavier than it would be if the earth were an isolated body in the heavens. Along the line between the centers of the earth and moon the tide-producing force is in the opposite direction to the earth's gravity and a body placed there is lighter. These differences in weight are quite small. A simple calculation (given in section A of the appendix to this chapter) shows that *F* is only about three millionths of the earth's gravitational force; *t* is even smaller.

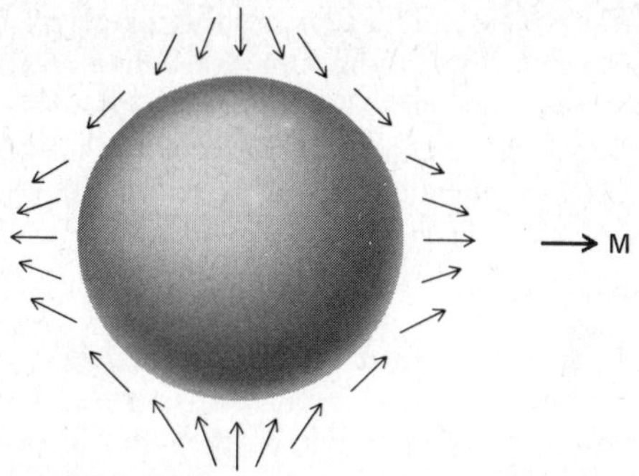

Figure 3.3 Direction of the tide-producing force at various places over the earth's surface.

Figure 3.3 shows that at most places on the earth's surface the tide-producing force, t, has horizontal components. These horizontal components of force acting on the oceans cause the flow of the tide. Figure 3.4a shows the magnitude and direction of the horizontal components of t. Due to this force there is a flow of water towards the moon on the right side of the diagram and away from the moon on the left. If we imagine the surface of the earth covered by a uniform layer of water, the action of the tide-producing force would produce the shape shown in Figure 3.4b. It is egg-shaped with its long axis in the direction of the moon.

The equilibrium theory of the tide assumes that the surface of the oceans actually takes the shape of the hypothetical fluid layer shown in Figure 3.4b. Now think of the earth as rotating from west to east about its axis while the axis of the tidal bulge continues to point in the direction of the moon. Any point on the surface of the earth is then going to pass through two places where the water is high and two places where it is low in the course of a day's rotation. Thus, there will be two high tides and two low

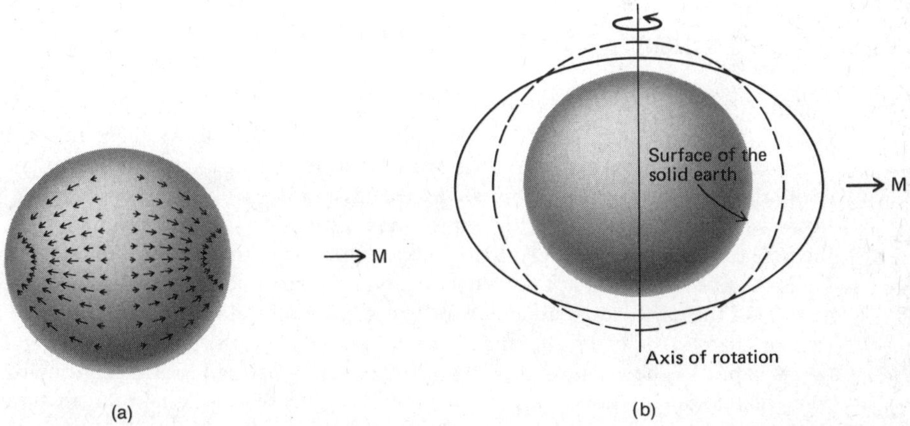

Figure 3.4 (a) the horizontal component of the tide-producing force. This is the force which moves the waters of the ocean so as to generate the tide. (b) The equilibrium tide which would result from the action of the force in Figure 3.4(a) on an ocean covering the entire earth. After the *Admiralty Manual of Tides,* HMSO, London, 1941.

tides each day. This is called a *semidiurnal* tide. The tide in most places is in fact semidiurnal; this simple theory thus accounts for the most obvious characteristic of the tide. Allowance should be made, however, for two additional factors. First, superimposed on the lunar tide there is an additional, smaller tide due to the sun. Second, the position of the moon relative to any place on the earth varies in a complex way because of the inclination of the earth's axis to the plane of the moon's orbit.

SOLAR TIDES

The solar tide-producing force is calculated in the same way as the lunar force. Now the mass of the sun is 27,100,000 times that of the moon and it is 389 times more distant from the earth. These differences make the solar tide-producing force only 0.460 times as large as the lunar force. The calculation of this ratio is shown in section B of the appendix to this chapter. The tidal effects of the other planets in the solar system are negligible.

The solar and lunar tides must be calculated separately and added to get the total tide. When the moon and the sun are in line the lunar and solar high waters coincide and the total tide is increased in the ratio 1.46 over the lunar tide alone. This is called the *spring tide*. (The name has nothing to do with the season of the year.) It is easy to see that spring tides fall on the dates of the new and the full moon, thus about every two weeks. When the moon is in quadrature (its first or third quarter) the solar low water coincides with the lunar high water and the tide is only 0.54 as great as the lunar tide alone. This is the *neap tide*. At some ports (Milford Haven on the south coast of Wales is an example) very large vessels can enter only on the spring tide. If the tide is missed a two week delay ensues. In an attempt to reach Milford Haven on the spring high water the *Torrey Canyon,* which was a few hours late, cut inside the light vessel off the Seven Stones. She grounded on Pollard Rock in one of the best known modern shipwrecks. The resultant oil spill caused grave damage to the English and French channel coasts.

THE MOON'S ORBIT

Figure 3.4b shows the axis of the earth perpendicular to the line between the centers of the earth and moon. This is a special case. Figure 3.5 shows the actual situation: The axis of rotation of the earth is inclined to the plane of revolution of the earth about the sun (the *ecliptic*). This is why we have seasons and why the days are longer in the summer than in winter; in winter the axis is inclined away from the sun and at the pole the sun is never seen at all. Only on the first day of spring and the first day of fall is the axis of rotation of the earth in a plane perpendicular to the line between the centers of the earth and sun, giving, on these two dates, day and night of equal length (the *equinoxes*). The altitude of the sun above or below the equator is its *declination* (see Chap. 2). The equatorial plane of the earth is inclined to the ecliptic at 23°27′. The greatest declination of the sun is then 23°27′ north (on the first day of summer in the northern hemisphere) or south (on the first day of winter).

The plane of the moon's orbit is inclined to the ecliptic at 5°8′, the position shown in Figure 3.6. Thus the moon may have a declination as great as $23°27' + 5°8' = 28°35'$ either north or south. Figure 3.4b applies only to the special

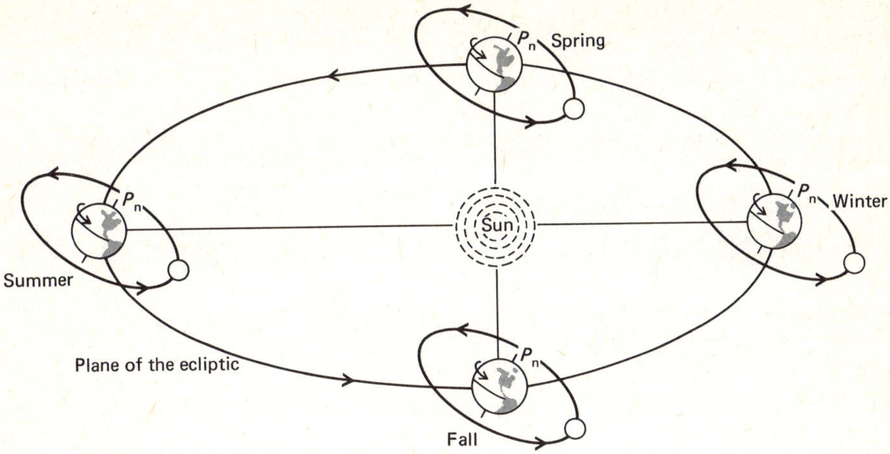

Figure 3.5 Relative positions in space of the orbits of the earth and moon. From Bowditch, *American Practical Navigation,* through courtesy U.S. Naval Hydrographic Office, 1958.

case when the moon's declination is zero. When the declination is not zero the equilibrium tide is inclined relative to the axis of rotation as shown in Figure 3.7. This has an important effect on the character of the tide. Consider a point at high latitude which follows the path AA′ as the earth rotates. It is easily seen from the drawing that this place passes through only one high water and one low water in the course of a day. At places in lower latitudes, such as B, the tide will be semidiurnal but the two high waters will not be of equal amplitude. This tide is shown in Figure 3.8, which follows directly from the study of Figure 3.7. It is the type of tide most commonly encountered.

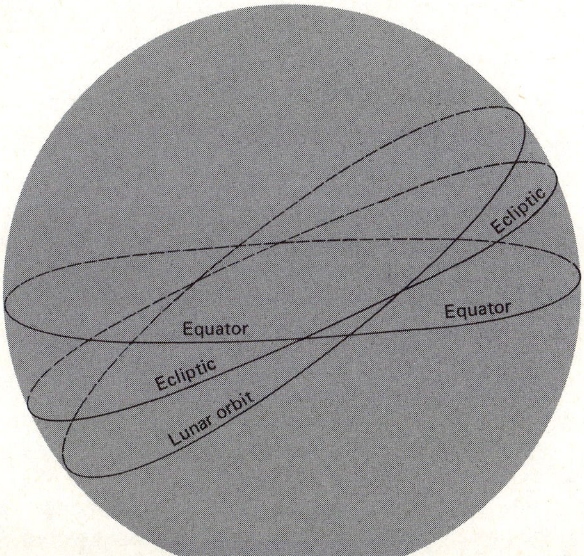

Figure 3.6 Relative positions of the equator, the plane of the ecliptic and the lunar orbit. After the *Admiralty Manual of Tides,* HMSO, London, 1941.

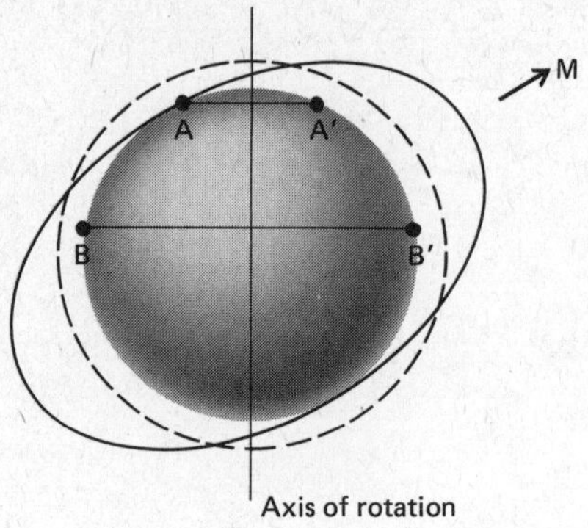

Figure 3.7 The equilibrium tide when the moon is at a declination different from zero. A point on the earth moving on the path AA′ as the earth rotates experiences a diurnal tide. Places at lower latitude, moving on a path such as BB′, will experience a semidiurnal tide with a diurnal inequality, as illustrated in Figure 3.8.

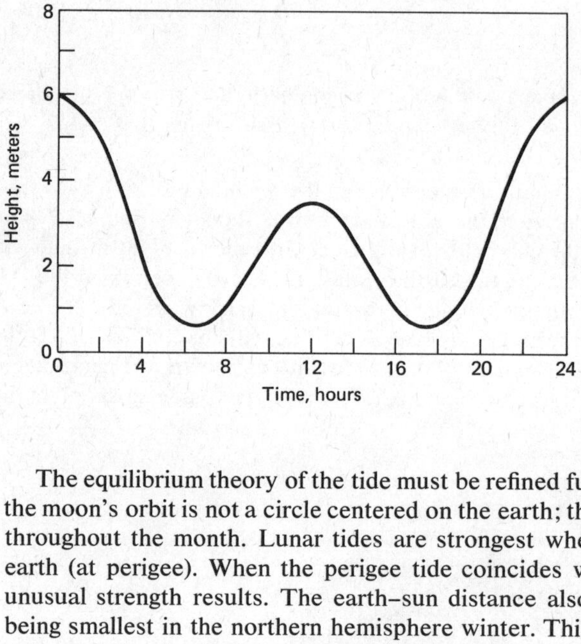

Figure 3.8 The tide for place B of Figure 3.7.

The equilibrium theory of the tide must be refined further to allow for the fact that the moon's orbit is not a circle centered on the earth; the earth–moon distance varies throughout the month. Lunar tides are strongest when the moon is closest to the earth (at perigee). When the perigee tide coincides with the spring tide, a tide of unusual strength results. The earth–sun distance also varies throughout the year, being smallest in the northern hemisphere winter. This introduces yet another periodic variation in the tide-producing force, in this case making the tide stronger in winter than in summer. Thus the tide-producing force is actually the sum of many periodic constituents each of which has its own characteristic pattern of variation. Despite these complexities, the tide-producing force and the equilibrium tide can be calculated from astronomical data for any time and place to any desired degree of accuracy. Does this mean that the actual state of the tide in the ocean is known to the same degree of accuracy?

THE EQUILIBRIUM TIDE AND THE ACTUAL TIDE

In order for the equilibrium tide actually to exist in the oceans, two conditions would have to be satisfied. First, the presence of the continental land masses would have to have no influence on the progress of the tide as the earth rotates. Second, the axis of

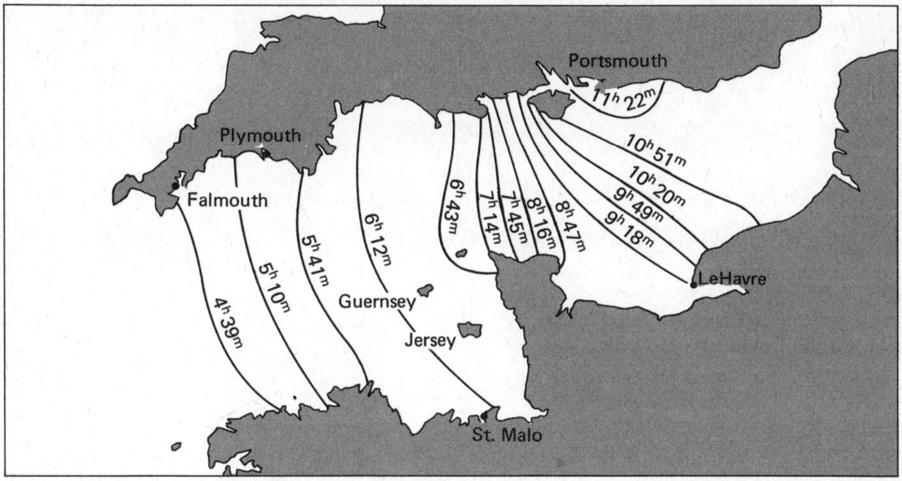

Figure 3.9 Progress of high water up the English Channel. The cotidal lines connect places where high water occurs simultaneously. Numbers give the interval between the time of the moon's transit of the Greenwich meridian and the time of high water. Note that the direction of advance of high water is the reverse of that predicted by the equilibrium theory. Based on BA chart 5058 with the sanction of the Controller, HM Stationery Office, and the Hydrographer of the Navy.

the equilibrium tidal bulge would have to continue to point along the line between the center of the moon and the center of the earth as the earth rotates on its axis. The equator is 21625 nautical miles in length and the lunar day is 24.84 mean solar hours. Dividing these shows that the equilibrium high water would have to travel at 871 knots to keep its place along the line between the centers of the earth and moon. In fact, it cannot travel nearly so fast. (The equation for the velocity of shallow water waves, developed in the next chapter, shows that this could occur only if the depths of the water over the earth were everywhere about 22,000 meters. In fact, the oceans are only about 4000 meters deep on the average.) Thus, the equilibrium tide cannot actually exist in the oceans.

Another way in which the ocean tide is quite different from the equilibrium tide is shown in Figure 3.9. Here are shown the *cotidal lines* for the English Channel. A *cotidal line* is a line drawn through all places where the time of high water is the same. In the chart, time is given relative to the time of the moon's transit of the meridian of Greenwich. Looking at the positions of the lines at successive intervals of time we see that the high water passes up the English Channel from Land's End to Dover. But this advancing wave of high water is not the bulge of the equilibrium tide, because it is advancing from west to east. The high water of the equilibrium tide must travel from east to west (since the earth rotates counterclockwise looking down on the north pole). The theoretical equilibrium tide does not exist in the oceans.

The calculated tide-producing forces shown in Figure 3.3 are correct. The real significance of the equilibrium tide is that it serves as a representation of these tide-producing forces. We have now to look more carefully at the physics of the response of the oceans to these forces.

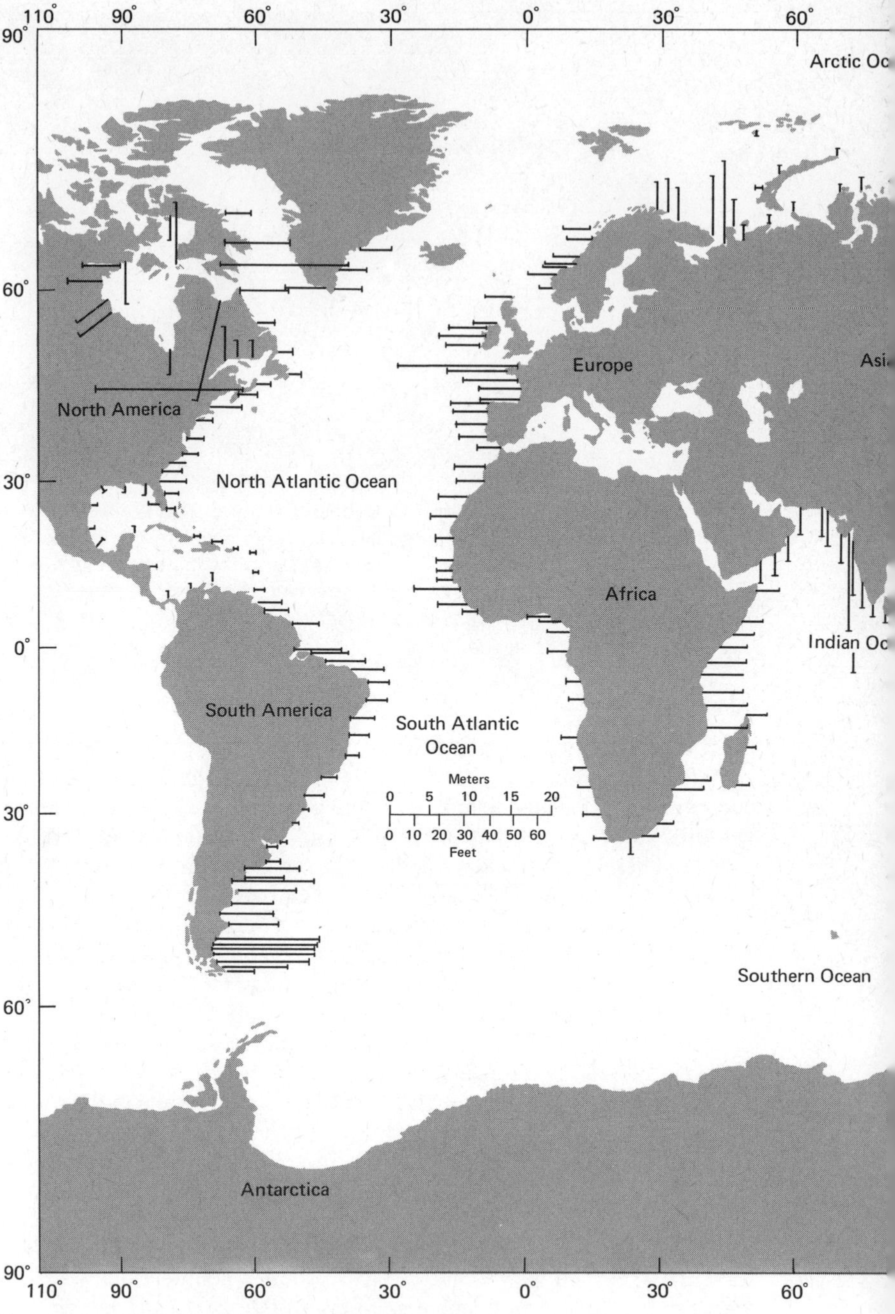

Figure 3.10 The observed height of the tide around the world.

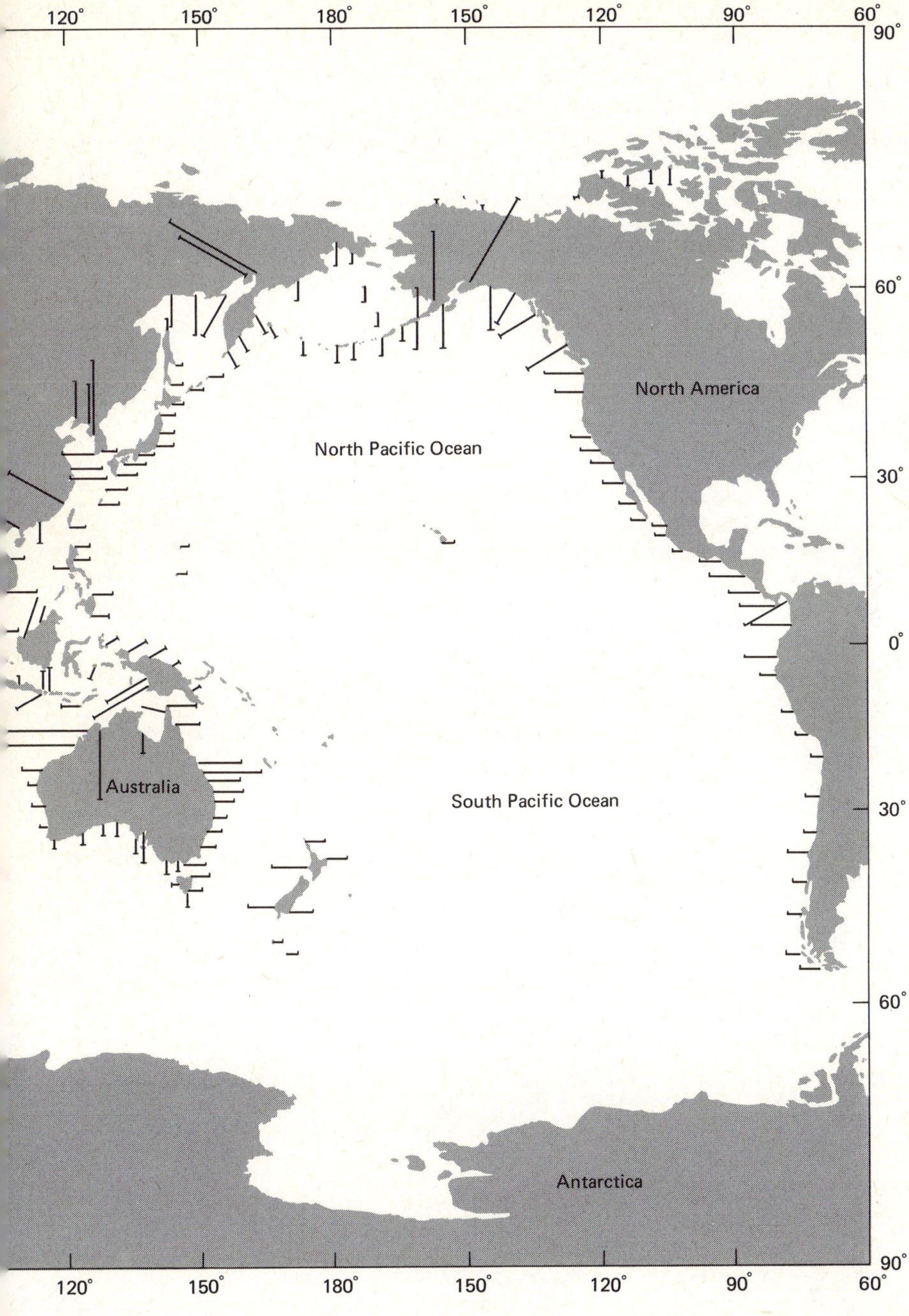
120°
150°
180°
150°
120°
90°
60°
90°
60°
30°
0°
30°
60°
90°
North America
North Pacific Ocean
South Pacific Ocean
Australia
Antarctica

RESONANT TIDES

In order to see how the oceans respond to the tide-producing forces it is helpful to look at Figure 3.10. This chart shows the world-wide distribution of the height of the tide, also called its *range*. It is evident that there is no significant tide in lakes or enclosed seas such as the Caspian or Mediterranean. The tide in the open ocean can be measured at islands such as Bermuda or Hawaii. It is always quite small, usually about 0.6 meters. It is only along coasts where the water becomes shallow and the shore irregular that a large tide occurs.

Consider next a nearly enclosed body of water such as Long Island Sound, Figure 3.11. If the Sound were walled off from the sea there would be no significant tide in it. Actually, the tidal range at the Race, the principal entrance, is only 0.7 meter but at Milford, well up the Sound, it has increased to 2.0 meters. The tide in the Sound is driven by the oceanic tide outside the Race but there is considerable amplification of the tide within the Sound. As the data in Figure 3.11 show, the tidal range in the Sound increases away from the principal entrance. It is easy to set up a simple hydraulic model with similar properties: a channel of rectangular cross section closed at one end and attached to a large tank of water at the other. Suppose that the water level in the large tank is oscillated up and down through a small range, slowly at first, then more rapidly. At a particular frequency the water in the attached channel starts to oscillate with a large amplitude. The pattern of oscillation is illustrated in Figure 3.12; it is called *resonant oscillation.*

The idea of mechanical resonance is a familiar one. For example, most musical instruments produce sounds by setting up a resonant oscillation; a piano string vibrates at its resonant frequency when struck, producing a pure tone. A little experimenting with a bathtub half filled with water will show that the water will slosh back and forth

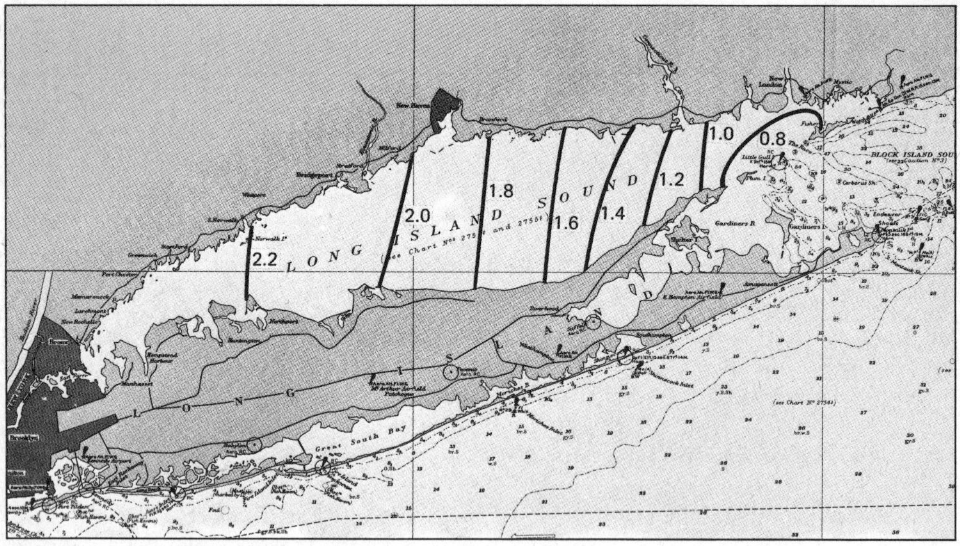

Figure 3.11 Amplitude in meters of mean high water in Long Island Sound. Base maps reproduced from BA chart number 2860 with the sanction of the Controller, HM Stationery Office, and of the Hydrographer of the Navy.

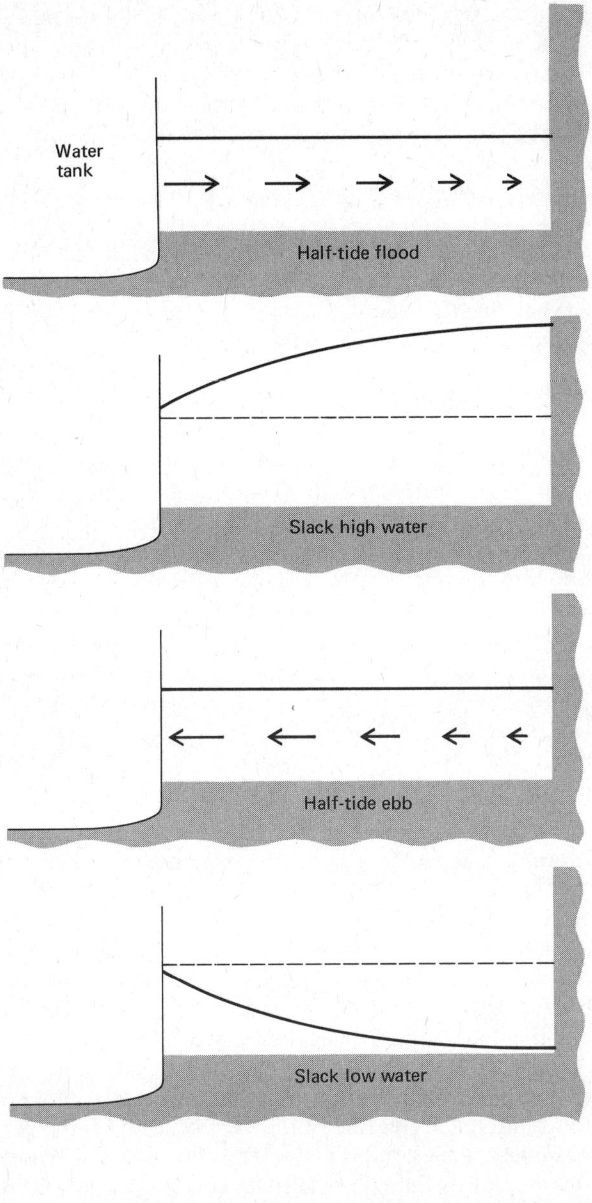

Figure 3.12 Oscillation of water level in a channel connected to a large water tank. At high and low water no current flows; at half-tide the water surface is level and the tidal currents are at a maximum.

in a regular way at one frequency only, its resonant frequency. In the resonant oscillation of water in the channel shown in Figure 3.12, high water, and later, low water, occur simultaneously at all places in the channel. Also, the range of oscillation of water level increases from near zero at the open end to a maximum at the closed end. The interval of time from high to low water and back to high water is the *period* of resonance, P. The resonant frequency is $1/P$. At high water and at low water no current is flowing anywhere in the channel; the tide is said to be *slack*. At half-tide, a quarter period after high or low water, the current in the channel is at its greatest strength. Of course, no current can flow at the closed end. The strongest current flows in and out of the entrance, the place where the amplitude is zero. These features correspond fairly closely with the actual tide in Long Island Sound. At the Race the tidal range is small but the tidal currents are very strong. At the western end of the Sound the tidal range is great and the currents weak.

The resonant period of water in a channel depends on its length and depth. The deeper the water and the longer the channel, the longer is the period of resonance (just as a long piano string sounds a lower note than a short one). The relationship between length, depth and period is easy to calculate. This is done in section C of the appendix to this chapter. There it is shown that for the period to be 12 hours, approximately the period of the semidiurnal tide-producing force, the length of the channel and the depth of the water must be related as in this table:

d (meters)	24	50	100	200
L (kilometers)	140	200	282	397

Long Island Sound is about 130 kilometers long and has an average depth of 22 meters. Its resonant period is therefore close to 12 hours; it is easily excited into large oscillations by a relatively small tide at its entrance. Wherever tides of very large amplitude occur, as in the Bay of Fundy or the Bristol Channel, it is because of resonant oscillation excited by the relatively weak tide of the open ocean.

Except in estuaries and small bays, the deflection of tidal currents by the Coriolis force results in a rotary motion superimposed on the oscillation of the water described above.

TIDE PREDICTION

The astronomical forces which cause the tide are known accurately and can be predicted far into the future, but as we have seen, the response of the sea to these forces varies greatly from place to place. Yet predictions of the tide are urgently required for the safe and efficient piloting of ships. How are they made? The solution is an example of a situation that arises frequently in applying physical analysis to a complex, natural phenomenon: Physical understanding is incomplete but in fact, a solution good enough for all practical purposes can be obtained by a combination of incomplete theory and empirical methods. The method that is used is to suppose that the tide at the place of interest follows the equilibrium tide that would be predicted for that place. It may have a different amplitude and may come at a different time than the equilibrium tide would, but the period of the tide must match the period of the driving force (as represented by the equilibrium tide). If the periods did not match there could be no fixed relation of force and response, a condition which must hold. (This principle was first formulated by Laplace in 1776 and first applied to tide prediction by Lord Kelvin in 1867.)

The method of prediction is as follows: Observations of the tide are made at the place of interest for a period of time, preferably 30 days or more. The regular amplitude and time differences between the observed tide and the calculated equilibrium tide are found from these data. Once these differences are known, predictions of the tide can be made for any given time by using the astronomically predicted positions of the sun and moon to establish the equilibrium tide at that time.

The accuracy of the prediction depends on the number of factors included in calculating the equilibrium tide. The simplest prediction considers only the lunar semidiurnal tide. The *lunitidal interval,* also called the *establishment of the port* is the time interval between the moon's transit of the meridian of the port and the next high water. Predictions based on the lunitidal interval neglect the diurnal tides and give no information about the tide height. More sophisticated predictions allow for the solar tide, changes in the moon's declination, and changes in the earth–moon distance. Predictions are then generated by adding these various terms. Until recently this was done mechanically with a tide-predicting machine, an array of cranks and pulleys connected so that their individual motions were added and indicated on a drum by a moving pen; now an electronic computer is used.

Predictions can be made to any degree of accuracy, but there is a practical limitation: Tides are significantly influenced by the weather. An onshore wind may, for example, retard the time of low water, and low barometric pressure makes the tide higher than would be expected. Thus, there is no point in trying to make predictions too fine. Typical practical accuracy attained is shown in these data taken at Liverpool over a period of one year:

Range of error in time of high water:	0–5	6–10	11–15	>15 min.
Percentage of predictions within range:	75	19	4	2

TIDAL POWER

Compared to all of the water used by hydroelectric plants, the volume of water set in motion by the tides along the coasts is simply immense. It has often been suggested that generators driven by the tides are a potential source of electric power free of the problems of air pollution that accompany the use of fossil fuels and the problems of radioactive waste disposal which will become a major limitation on the large-scale use of nuclear power.

Two practical difficulties stand in the way of the utilization of tidal power. First, the cyclic character of the tides means that steady power generation is not possible unless the tidal power facility contains a vast water storage system which can be utilized to supply a steady flow of waters to the powerhouse turbines. Second, the height of the tide is in most places insufficient to give the head (pressure due to height of water) required for the efficient operation of turbines of reasonable size. The power that can be produced by flowing water depends on the product of the volume of water moving and the water pressure. The best site for a hydroelectric installation is one where the head can be made large, as in a river valley where a high dam can be built. The greater the head, the smaller the volume of water flow required for a given amount of power. And the smaller the volume of water, the smaller the size (and hence the cost) of the turbines required. In a tidal power installation the average available head must be less than the tidal range; a large and expensive set of turbines is required to utilize the large volume of low pressure water set in motion by the tide. These two factors, cyclic flow and low head, conspire to make the initial cost of a tidal power installation very large. Since interest charges on the initial investment are part of the cost of generating power, tidal power costs tend to remain above those of coal-fired and atomic power plants even at the most favorable sites.

Moreover, the number of favorable sites is very limited. Figure 3.10 shows that tides greater than 10 meters occur at only a few places. One site on the English Channel has already been utilized (the Rance estuary on the Brittany coast). A proposal to build a tidal power plant using Passamaquoddy Bay, near Eastport, Maine, as a reservoir has been under discussion on and off for nearly thirty years. The capital cost has always proved to be much too high. Other possible sites, as on the Patagonian coast, are too remote to be of much use. But even if all available sites were utilized, the power generated would be only 64,000 megawatts, small compared to the 734,000 megawatts of electrical generating capacity in the world in 1964. Tidal energy, while it may be locally useful, is not going to supply the growing demand for electric power in the industrialized nations.

BORES, OVERFALLS, RIPS, AND RACES

Where the range of the tide is great or the tidal currents strong, the ebb and flood of the tide may be violent in character, sometimes creating dangerous water conditions such as the maelstrom which was thought to have engulfed Captain Nemo's *Nautilus*. A tidal *bore* is the advance of a nearly vertical wall of turbulent water up a river or estuary on the flood tide. *Overfalls, rips* and *races* occur when a swift tidal current runs over an irregular bottom; the irregularities in the bottom are mirrored in corresponding irregularities in the surface of the flowing water.

An elementary theory for the speed of advance of a tidal bore is easily constructed; it is a good example of how a natural phenomenon, sufficiently simplified, can be understood by physical analysis. The theory gives the velocity of advance of the bore in terms of the depth of the river channel. (The problem of the original formation of the bore, which involves transient rather than steady conditions, is not so easily dealt with.) To find the velocity of a bore, consider a channel of uniform width and depth. Let it be one unit wide. Because the bore is moving relative to the channel it is difficult to construct equations describing the flow of water; but an equivalent problem which is easy to solve can be made up. Imagine a current of velocity v, flowing to the left at such a speed that the bore, which is advancing to the right, is held stationary in front of us. This is illustrated in Figure 3.13. Clearly, v must be of the same magnitude, but opposite direction, as the velocity of the bore upstream. If the magnitude of v can be found, the problem of the velocity of advance of the bore is solved. To calculate v we make use of two principles of mechanics, the law of conservation of mass (which may be called the equation of continuity when dealing with a

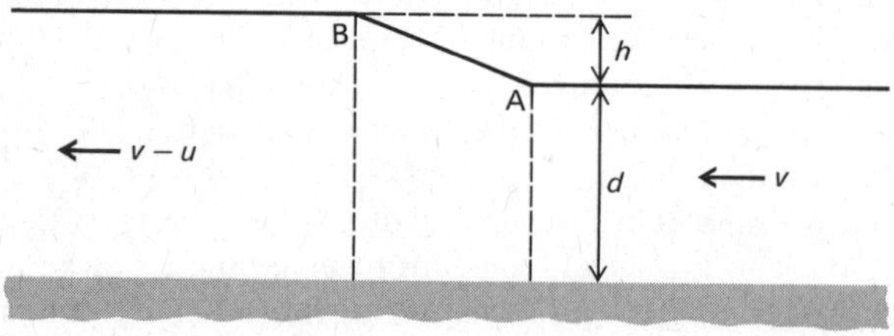

Figure 3.13 A bore of velocity *v*, which would advance to the right into still water, held stationary by the current *v* flowing to the left.

fluid) and the law of conservation of energy. To keep the algebra simple, we will assume that the height of the bore is small compared to the depth of the water in the channel. To apply conservation of energy, consider a particle of mass m moving at the surface of the stream from the right. As it encounters the bore its elevation is increased by the height of the bore, h, which means that its potential energy is greater by mgh (where g is the acceleration due to gravity; the particle has been moved a distance h against the force mg). Because energy is conserved, there must be a corresponding decrease in the particle's kinetic energy—its velocity to the left of the bore must be lower. This decrease occurs because there is a current u flowing upstream behind the bore to supply the water required for the additional height of the water level. The particle velocity on the left side of the bore is then $v - u$. Since the kinetic energy of a particle of mass m moving with velocity v is $\frac{1}{2}mv^2$, it follows that

$$\tfrac{1}{2}mv^2 - \tfrac{1}{2}m(v-u)^2 = mgh \tag{3.2}$$

Continuity (conservation of mass) requires that as much water leave the cross section at B as enters at A. Remembering that the channel is of unit width, the area at B is $d + h$ while that at A is d. In unit time a volume vd enters at A and a volume $(d + h)$ $(v - u)$ leaves at B. Thus

$$vd = (d+h)(v-u) = vd - ud + hv - hu \tag{3.3}$$

or

$$hv = u(d+h) \tag{3.4}$$

From Eq. 3.2, on multiplying out terms,

$$\tfrac{1}{2}mv^2 - \tfrac{1}{2}mv^2 + mvu - \tfrac{1}{2}mu^2 = mgh \tag{3.5}$$

or

$$uv - \tfrac{1}{2}u^2 = gh \tag{3.6}$$

We now utilize the assumption that h is much smaller than d. Since the channel is nearly the same depth before and behind the bore, there is a large cross-sectional area in which to bring in the extra water required, which means that u is much smaller than v. Then u^2 is so small in comparison with uv that it may be neglected, and Eq. 3.6 becomes $uv = gh$. Now eliminate u with Eq. 3.4 and rearrange terms to find that

$$v = \sqrt{g(d+h)} \tag{3.7}$$

Thus the velocity of advance of the bore depends only on the depth of the water behind its advance. The deeper the water the greater the velocity. To illustrate the magnitudes involved, a bore advancing so as to produce water 2 meters deep has a speed of 21 kilometers per hour. Clearly a person standing near such a channel might have difficulty escaping a tidal bore and, in fact, a number of cases of persons swept into the rising waters of a bore have been recorded. Fortunately, the bore announces its arrival with a roar to be heard some distance off. Our analysis also explains why the observed height of a bore is usually greater near the edge than in the center of the channel. In most rivers the channel shoals near the edges, which causes a reduction in the speed of the bore. To compensate for the shoal edges, water is transferred from the center of the stream to the edges, building up the height of the bore (increasing h to compensate for decreasing d in Eq. 3.7) sufficiently to keep a uniform velocity of advance across its front.

A simple theory of a tidal race may be constructed in much the same way as the

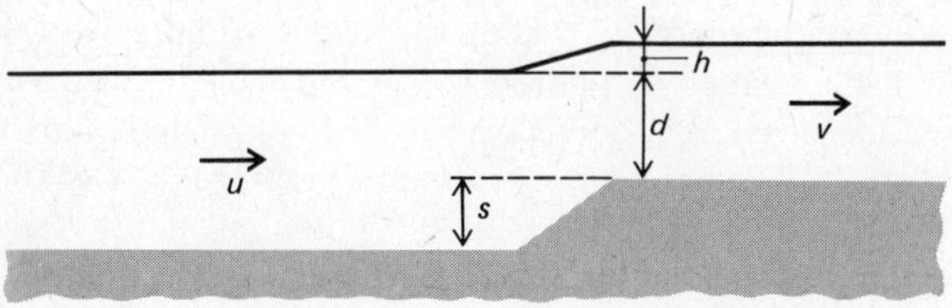

Figure 3.14 Water flowing over a step of height *s* in the floor of a channel.

theory of the tidal bore. In a race a current runs over an uneven bottom, as in Figure 3.14, causing a change in the elevation at the surface. The equations of continuity and conservation of energy are applied in the same way as before (try the derivation if you like) to find, when h and s are small compared to d, that

$$h = \frac{s}{1 - (gd/u^2)} \tag{3.8}$$

If $u^2 > gd$, then $h > 0$ and the bottom topography is mirrored positively on the surface, as in Figure 3.14. As u increases h becomes smaller and, for sufficiently high velocities, all but disappears.

If $u^2 < gd$, then $h < 0$. In such a case, the shape of the race is not as in Figure 3.14; instead there is a depression of the surface as the water runs over the step on the bottom. This case is shown in Figure 3.15. The shape of the bottom will be mirrored in reverse on the surface with a magnification ratio $h/s = 1/[1 - (gd/u^2)]$. For small velocities h/s is small. As u increases, h becomes greater. As u^2 approaches gd, h becomes large without bound, but we have then passed beyond the range of conditions (h and s are small compared to d), under which Eq. 3.8 was derived. Clearly, however, a race or rip of great strength can result from the proper combination of d and u; if the water becomes highly turbulent in the rip, an overfall results. All of these situations can be substantial hazards to the management of ships and boats.

TIDAL FRICTION

The ebb and flood of the tide moves vast quantities of water about over the earth's surface. Because of the viscosity (resistance to flow) of water and the friction of tidal currents over the ocean bottom, energy is dissipated into heat by this movement. The principal source of the energy dissipated is the orbital energy of the moon. The friction of the oceanic tides therefore exerts a subtle influence on the moon's motion; it accelerates the moon's orbital velocity. If the tide moved without friction, the tidal bulge shown in Figure 3.16 would remain directly in line with the moon. Because

Figure 3.15 Formation of a tide rip for the case where the water velocity is $u^2 < gd$.

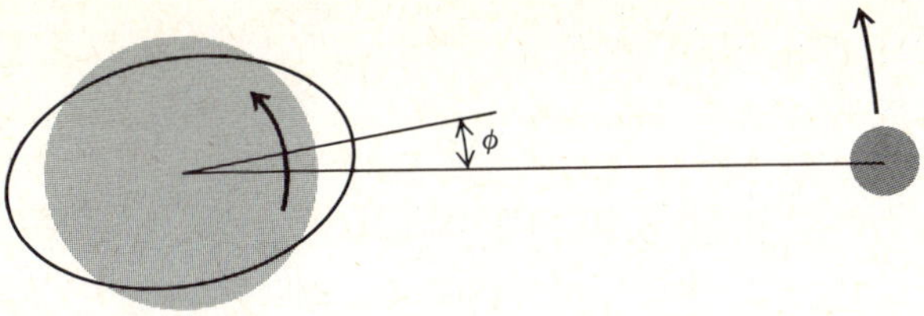

Figure 3.16 Due to friction between the rotating earth and ocean water, the tidal bulge on the earth is carried ahead of the position of the moon by the angle ϕ.

there is friction and because the earth rotates faster than the moon revolves, the tidal bulge is carried ahead of its proper position by the amount ϕ. The unsymmetrical distribution of mass on the earth relative to the moon then exerts a torque (twisting force) on the moon. This torque tends to pull the moon ahead in its orbit and, simultaneously, to slow down the rate of rotation of the earth. Astronomical observations show that the moon is experiencing a constant acceleration of $22''.4$ century^{-2} = 1.09×10^{-23} rad sec^{-2}. The increasing orbital velocity of the moon means that the size of its orbit is increasing. The rate of increase is 1.0×10^{-7} cm sec^{-1}.

Modern astronomical observations demonstrate the existence of a significant tidal friction; there is also evidence of the effect of this friction in the historical record. A number of eclipses observed in ancient times appear to have occurred at the wrong place by amounts which can be accounted for by the tide-induced change in the length of the day (over many centuries a small change in rotation rate produces a significant difference in longitude). Evidence from early geologic times can be obtained from the study of fossil corals and clam shells: A clam living in an intertidal zone grows only when covered by water. Its shell develops growth rings which are a record of successive high waters. On these rings the sequence of spring and neap tides is revealed. By counting rings the number of days between springs, and hence the number of days in the lunar month, can be determined. Such observations indicate that over the past 450 million years the month has lost about two days due to slowing of the earth. It is interesting to speculate on the great tides which must have existed in the past geologic ages when much more of the earth was covered with shallow seas and the moon was closer (remember, the tide-generating force depends on $1/r^3$).

And what of the future? The rotation rate of the earth will continue to decrease, the revolution of the moon in its orbit about the earth, to speed up. The moon, because of its faster revolution, will recede further from the earth. The ultimate effect of tidal friction in the earth-moon system will be that the length of the month and the length of the day will become about the same (about 50 of our present days), the moon will remain over one place on the earth, and be about 40 percent further away than it is now. The lunar tides on the earth will cease. If the earth and moon were an isolated system this is all that would happen. But they are not. Solar tides will still act on the earth and, through tidal friction, will continue to slow the earth's rotation. The moon will then appear to move across the sky in the reverse from its present direction and its revolution will be slowed as the lunar tide is reborn. Slower revolution means a decrease in the earth–moon distance with a consequent strengthening of the tide. As the moon spirals closer to the earth the tidal forces on it will eventually exceed its strength and the moon will be shattered to fragments. Thus the tidal friction means, over a span of billions of years, the demise of the moon itself.

APPENDIX

A. Magnitude of the Moon's Gravitational Force at the Surface of the Earth (*F* in Fig. 3.2)

A unit mass on the earth's surface is attracted to the earth by a force whose magnitude is g, the acceleration due to gravity. (This follows from Newton's second law of motion, $F = ma$, with $m = 1$ and $a = g$.) The gravitational force on this same unit mass due to the moon's gravity is much less, because first, the mass of the moon is much less than the mass of the earth, and second, the moon is much further away. How much less may be calculated by ratios. The mass of the earth, M, produces a force equal to g on a unit mass a distance of r, the earth's radius, from its center. The mass of the moon, m, a distance s away then produces a force equal to

$$F = g\,\frac{m}{M}\,\frac{r^2}{s^2}$$

Since $m/M = 81.53$ and $r/s \simeq 60$,[1] it follows that $F = 3.3 \times 10^{-6}g$.

B. Magnitude of the Solar Tide

According to Eq. 3.1, the gravitational force falls off with distance as $1/r^2$. However, the tide-producing force, t in Figure 3.3, depends on the difference between gravitational forces at different places on earth. Consequently, it decreases even more rapidly with distance. This may be seen by calculating the difference in gravitational force, F, between bodies of masses M and m when the distance between these is increased from r_1 to the slightly larger distance r_2. Thus,

$$F_2 - F_1 = G\,\frac{mM}{r_2^2} - G\,\frac{mM}{r_1^2} = GmM\left(\frac{1}{r_2^2} - \frac{1}{r_1^2}\right)$$

$$= GmM\left(\frac{r_2}{r_2^3} - \frac{r_1}{r_1^3}\right) \simeq G\,\frac{mM}{r^3}\,(r_2 - r_1)$$

The approximation holds because $r_1 \simeq r_2 = r$. The tide-producing force decreases with distances as $1/r^3$.

The mass of the sun is 27,100,000 times that of the moon and it is 389 times more distant from the earth. The relative magnitude of the solar tide-producing force is then

$$\frac{27{,}100{,}000}{(389)^3} = 0.460$$

Despite the very much greater mass of the sun compared to the moon, the earth–sun distance is so much larger than the earth–moon distance that the solar tide-producing force is less than half of the lunar force.

[1] The symbol "≃" means "approximately equal to."

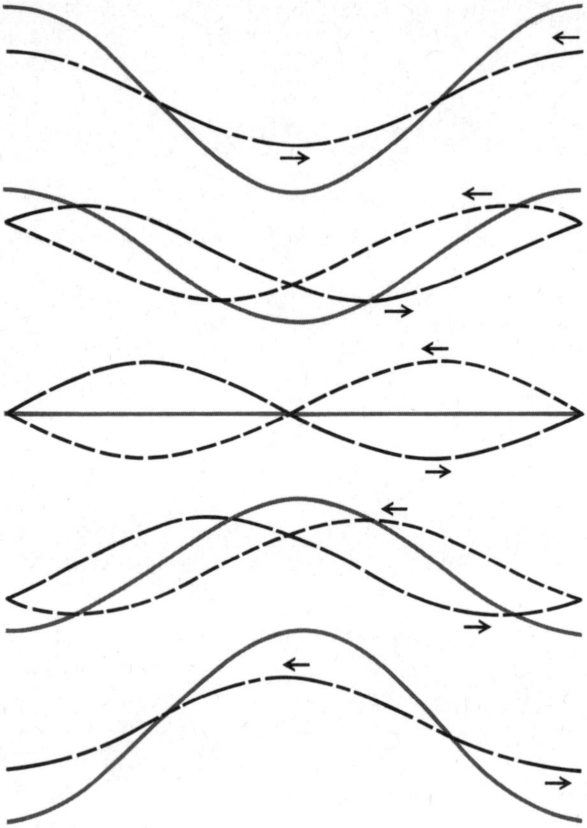

Figure 3.17 A standing wave (solid line) can be considered to be the sum of two traveling waves (dashed lines) of the same wavelength but moving in opposite directions.

C. Resonant Tides

The resonant period of the water in the channel discussed on p. 63 depends on the channel length and the water depth. The motion of the fluid shown in Figure 3.12 is a form of wave motion which may be described as a *standing wave* (to distinguish it from a *traveling wave* where the crests and troughs move regularly forward). The standing wave may be thought of as the result of two traveling waves of equal amplitude moving in opposite directions (Fig. 3.17). At high or low water the troughs and crests of the two traveling waves coincide, whereas at halftide they are exactly out of phase. For the standing wave shown in Figure 3.12 the wave length, λ, the distance between crests, is equal to $4L$ (L is the length of the channel). It will be shown in the next chapter that the velocity of a wave whose wavelength is long compared to the depth of the water, h (a very good approximation for all waves generated by the tide) is $v = \sqrt{gh}$. Since the velocity v, period P, and wavelength λ of any wave are related by $v = \lambda/P$, the resonant period of a gulf closed at one end is

$$P = \frac{4L}{\sqrt{gh}} \qquad (3.9)$$

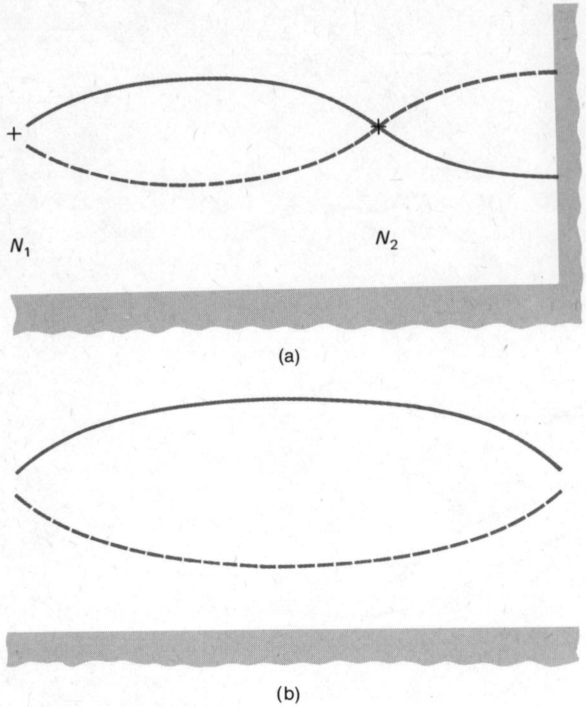

Figure 3.18 (a) Standing oscillation in a gulf whose length is three quarter wavelengths. Nodes occur at N_1 and N_2. (b) Standing oscillation in a strait or sea open at both ends.

Resonances other than those involving a single node (place where the amplitude of oscillation is zero) are possible in partially enclosed bodies of water closed at one end. Two or more nodes may be present; it is only necessary that L be an integral multiple of $\lambda/4$, as in Figure 3.18a. Resonance may also occur in a channel open to the sea at both ends as in Figure 3.18b; in this case $P = 2L/\sqrt{gh}$, since the node must be across the center of the channel. Oscillation of this kind occurs in the Straits of Magellan.

Further Readings

(Arranged in order from elementary to advanced.)

CLANCY, E. P., *The Tides*. New York: Doubleday, 1968.

TRICKER, R. A. R., *Bores, Breakers, Waves and Wakes*. New York: Elsevier, 1964.

MACMILLAN, D. H., *Tides*. New York: Elsevier, 1966.

DOODSON, A. T., and WARBURG, H. D., *Admiralty Manual of Tides*. Hydrographic Department, Admiralty, London, 1941.

DEFANT, A., *Physical Oceanography*, Vol. II. New York: Macmillan, 1961.

4 Interfaces I: Waves and Surf

The tide-producing force is a *body force*: It acts on every particle within the volume under consideration. The next two chapters are concerned with the response of the earth to forces acting at surfaces, the forces produced by the wind blowing over the ocean, for example. Interfaces are a special concern because we live at the air-land or the air-water interface; it is our natural environment. We will look at the action of the wind in exciting ocean waves and in blowing away the surface of the land, and at the shaping of the face of the continents by running water.

WATER WAVES

Look at a small pond on a day when there is no wind. The surface is as smooth as a mirror, and the reflections are perfect. But only a very slight breeze need blow over the pond to set its surface in motion: Waves are then generated and move in the direction of the wind.

Stand by the ocean on a day when there is no wind whatsoever. The water surface will never be truly at rest; waves will always be arriving at the shore. These are wind-generated waves also, but the storm which was their source may have been a thousand miles or more away. A fundamental characteristic of waves is that they transport energy over long distances without transporting matter.

The size of the waves generated on a pond is not likely to be of much consequence even in a stiff gale. At sea it is otherwise. When a strong wind acts over a large expanse of water, waves of great height can build up (Fig. 4.1). Lieutenant F. C. Margraff, watch officer of the *USS Ramapo,* once measured a wave 34 meters high as his ship navigated through an intense storm in the Pacific. Waves of 20 meter height have been reported by many other reliable observers at sea. The management of a ship in such seas tests both men and equipment to their limit and indeed, sometimes beyond.

Driving head-on into waves with the bow and stern supported as in Figure 4.2, for example, the hull of a ship must carry the whole weight of boilers and engines at its center. Design of a structure to sustain such loads remains a challenge, in large part

Figure 4.1 The North Atlantic in winter; the size of the waves may be judged from the length of the ship, H.M.S. *Renown,* 228 meters. Courtesy of the National Maritime Museum.

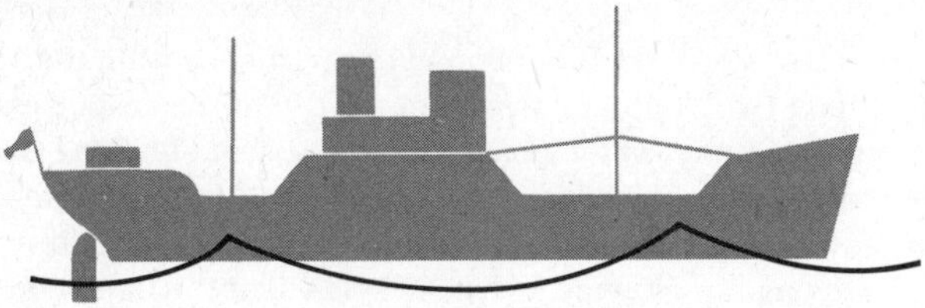

Figure 4.2 When the distance between wave crests is about equal to the length of the ship, the hull may have to carry the entire weight of the propulsive machinery without support except at the ends.

because it is difficult to describe adequately in a quantitative way the action of a storm-tossed sea. Even the representation of such a sea on canvas is often a vexing problem for the artist.

High waves often threaten the mariner with another danger. When wave amplitude is high there is a tendency for the water at a wave crest to move forward faster than the wave itself. The wave then *breaks;* a stream of turbulent water flows down the front slope of the wave, as shown in Figure 4.3. This flow of turbulent water is capable of exerting large forces on structures it encounters. It can easily sweep the deck of a ship clean of all movable objects. The sweep of water across the spar deck

Figure 4.3 A Coast Guard rescue vessel 16 meters long passing through a breaking wave on the Oregon coast. Courtesy U.P.I.

is one of the greatest hazards faced by the crew of a sailing vessel on an ocean voyage. Approaching shore, waves always tend to break as they enter the shallow water. The turbulent water of breakers of modest size contributes to the excitement of surf swimming and to the formation of beaches; in storms breakers are responsible for the destruction of much seashore property.

PROPERTIES OF WAVES

Since, as we have observed before, large waves may arrive on shore on a windless day, waves evidently have the capacity of transmitting energy over long distances. They do not transport matter; if on a calm day one observes the motion of a float upon the water, as waves pass, the float moves up and down, and to and fro, but after the passage of many waves the average position of the float remains essentially unchanged. Only very slowly will the float advance in the direction in which the waves are passing. One can observe another characteristic of waves when one swims in clear water at a beach with a shallow slope. Dive into the trough in front of an advancing wave and come to rest just below the surface in a place where a mark, such as a stone on the bottom, can be watched. As the wave arrives the swimmer will be pulled sharply across the bottom towards the wave and then returned equally sharply to his original position. Strong horizontal currents accompany the passage of the wave. But if one dives well below the surface in deep water, no water motion at all will be felt as waves pass. The crew of a submarine can sit in comfort while a ship above is being tossed about by a heavy sea. Thus, wave action is confined to a zone near the surface.

A short period of study of the sea surface will suffice to convince an observer of the complexity of water waves. Often no regular pattern of motion can be discerned; crests and troughs appear and vanish in a pattern which is almost random (Fig. 4.4).

Figure 4.4 Often the dispair of the marine artist because of its irregularity, the sea surface commonly shows little detectable pattern.

Little of the regularity characteristic of tides is seen. Indeed, as Lord Rayleigh observed, "The basic law of the seaway is the apparent lack of any law," so there are going to be complications in the analysis of ocean waves. The easiest way to proceed is to analyze the properties of a simple train of waves. An approximation to a true sea state such as that in Figure 4.4 can then be built up by the superposition of a number of independent wave trains.

A wave train is illustrated in Figure 4.5. Figure 4.5a is constructed by plotting the elevation of the water surface at every point as observed at a given instant; the graph is thus elevation, y, against distance, x, along the wave train. The position of the waves at a later time is shown by the grey line. Whenever the distance λ, called the *wavelength* is traversed, the wave pattern is exactly repeated. Figure 4.5b is constructed by plotting the elevation of the water surface at every moment of time as observed at a given point; the graph is thus elevation against time. Whenever the interval of time P called the *period,* elapses, the wave pattern is exactly repeated. The number of waves passing a point in unit time is $1/P = f$, the *frequency* of the wave. Since in unit time f wave crests pass a point, and the distance between crests is λ, it follows that the velocity of the wave, that is, the speed of advance of any wave crest or trough, is

$$v = f\lambda \tag{4.1}$$

The starting point in the analysis of wave motion is the determination of this velocity.

In principle the analysis of wave motion on the surface of water is a straightforward problem. For motions of the scale of interest in ocean waves, water may be considered to be an incompressible fluid with no viscosity or surface tension. The motion of its surface is governed by Newton's second law and the law of conservation of mass. In hydrodynamics these are expressed as the *equation of motion* and the *equation of continuity*. These equations must be solved in order to calculate the motion of the water surface as waves pass. The difficulty which arises is that the equation of motion for this problem is nonlinear; that is, it contains the products of unknown terms. There is no known way of solving it exactly. Solutions must be found by making successive approximations. Rather than following this formal mathematical procedure, it will be easier for us to examine two particular, but important,

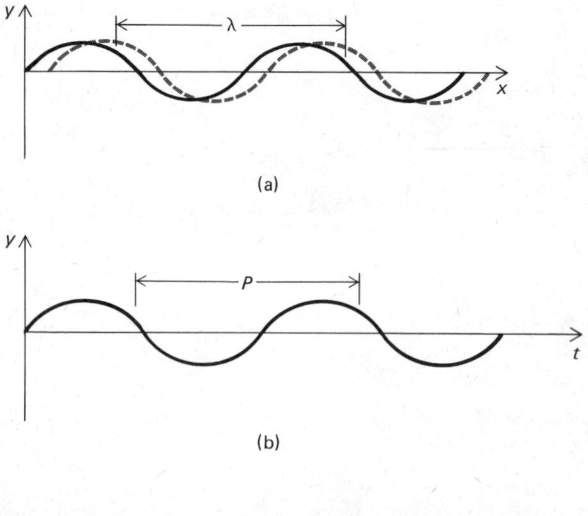

Figure 4.5 Wave trains in space (a) and time (b). The wavelength is λ and the period, P. The dashed line in (a) Shows the position of the wave at a slightly later time.

classes of water waves where, by simple physical arguments, solutions can be obtained. The properties of more general waves can then be inferred with some confidence. The two simple cases are waves of small amplitude, first in shallow water and second in deep water.

SHALLOW WATER WAVES

In the study of wave motion water may be considered shallow if its depth is small compared to the wavelength. Waves approaching a beach inevitably become shallow water waves. The wave associated with the advance of the semidiurnal tide is a shallow water wave even in the deepest parts of the ocean because of its great wavelength. In the passage of a wave in shallow water the vertical component of the velocity of a particle is negligible compared to the horizontal component; a particle moves forward and back as the wave passes.

In Chapter 3 the velocity of a tidal bore advancing up a channel was calculated. The bore originated because the water level at the end of the channel was suddenly raised. A *negative* bore is formed if the water level at the end of the channel is suddenly lowered. Positive and negative bores are compared in Figure 4.6. As in the analysis of the positive bore in the last chapter, a counter-current velocity, v, is chosen so that each bore remains stationary in our frame of reference. If the passage of a positive bore is followed by the passage of a negative bore up the channel, then an approximation to a wave is generated. The calculation of the speed of the negative bore is done in the same way as that for a positive bore. The equations for conservation of energy in the two cases are

$$\tfrac{1}{2}mv^2 - \tfrac{1}{2}m(v-u)^2 = mgh \qquad (3.2) \qquad\qquad \tfrac{1}{2}m(v+u)^2 - \tfrac{1}{2}mv^2 = mgh \qquad (4.2)$$

The corresponding equations for conservation of mass are

$$vd = (d+h)(v-u) \qquad (3.3) \qquad\qquad vd = (d-h)(v+u) \qquad (4.3)$$

Equations 4.2 and 4.3 are rearranged and combined by the same steps as were used with Eqs. 3.2 and 3.3 to find that in the case where u is much less than ($\ll$) v,

$$v = \sqrt{g(d+h)} \qquad (3.7) \qquad\qquad v = \sqrt{g(d-h)} \qquad (4.4)$$

In each case the velocity of the bore is determined by the depth of water (as increased or reduced by the bore itself). Note that if the amplitude of the bore is small compared to the depth, then both bores propagate with almost the same velocity, $v = \sqrt{gd}$.

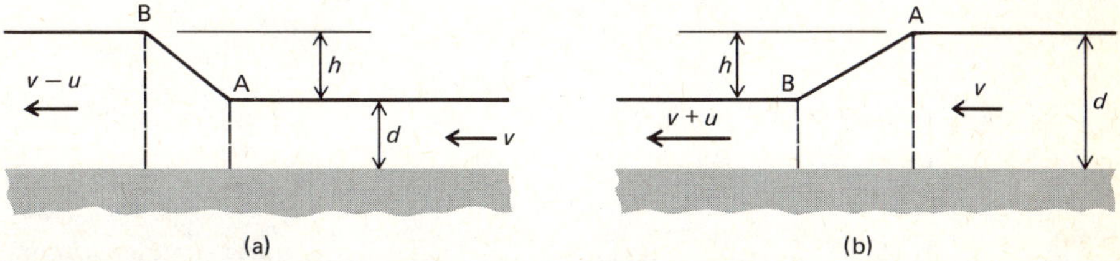

Figure 4.6 Comparison of positive (a) and negative (b) bores. The hypothetical counter-current velocity v will hold the bore stationary in front of the observer.

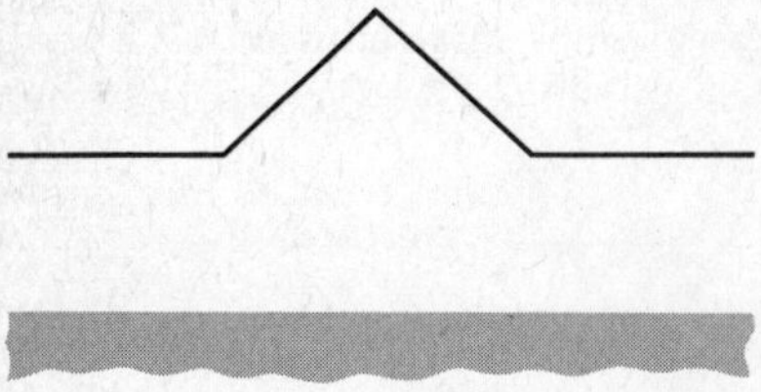

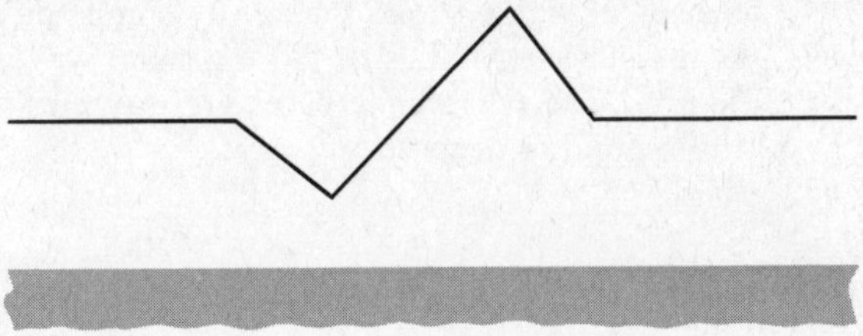

Figure 4.7 (*Left.*) Passage of a positive bore followed directly by a negative bore is equivalent to a half-wave passing up-channel.

Figure 4.8 (*Right.*) Two positive and two negative bores model a complete wave.

How is it possible that a positive bore, say, can travel up a channel at a speed much faster than the speed of the water flowing behind it, that u can be very much less than v? This is because the height of the bore is reasonably small compared to the channel depth. The whole depth of the channel is used to supply the additional water required to raise the water level; consequently, the velocity of flow required to supply this water remains fairly low; a high wave speed can result from only a small speed of the individual water particles.

If the passage of a positive bore up a channel is followed directly by passage of a negative bore, something approximating a wave has been generated (Fig. 4.7). A better approximation can be made by using two positive and two negative bores as in Figure 4.8. When $h \ll d$ the positive and negative bores travel with the same speed so that the whole pattern will stay together and the wave will retain its shape. The pattern can be repeated any number of times so as to generate a wave train.

This analysis immediately reveals the important properties of shallow water waves. First, the velocity of a wave, $v = \sqrt{gd}$, depends on the depth of water and is independent of wave period and amplitude (so long as $h \ll d$). Second, the velocity of water particles is small compared to the wave speed. The pattern of particle motion is shown in Figure 4.9. This is the pattern revealed by swimming under the wave approaching a beach. The water is in motion right down to the bottom. Third, as waves enter shallower and shallower water, as in advancing up a beach, $h \ll d$ no longer holds, the crests, moving at $v = \sqrt{g(d+h)}$, travel faster than the troughs, moving at $v = \sqrt{g(d-h)}$. Crests then begin to advance upon the troughs, (Fig. 4.10), the front face of the advancing wave steepens, and eventually, a crest overtakes the trough ahead of it and the wave breaks.

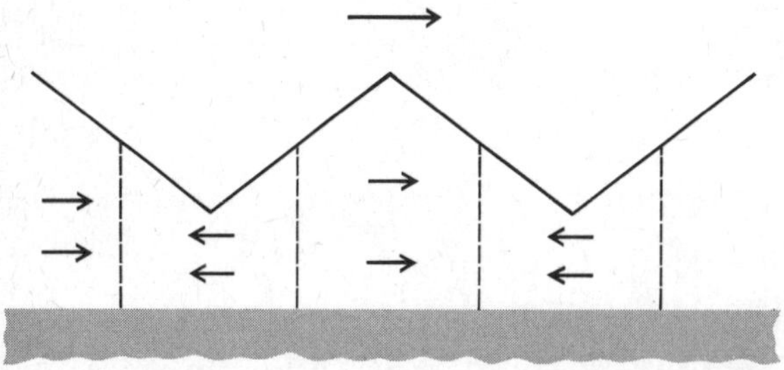

Figure 4.9 Direction of particle motion during passage of waves in shallow water.

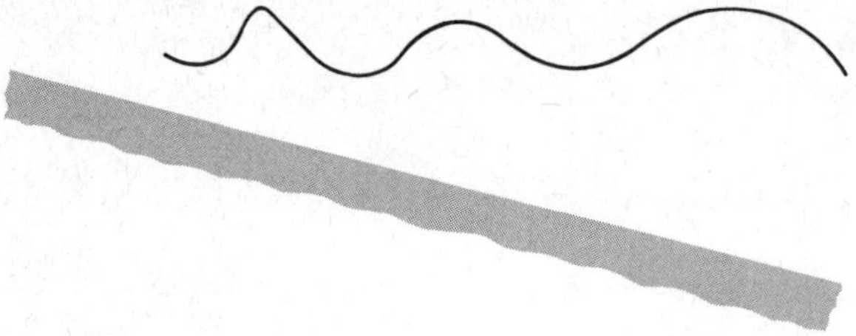

Figure 4.10 On advancing into shallow water the front face of the wave crest steepens while the back face becomes flatter. Continued steepening of the front face results in breaking of the wave.

DEEP WATER WAVES

The shallow water approximation accounts for the characteristic features of waves advancing upon a beach. The approximation fails in deep water because significant motion of the water particles does not occur at a depth greater than a few wavelengths. Thus, in deep water, the equation of continuity is not so easily applied (because it is no longer possible to construct a hypothetical channel of known dimensions through which the water must flow in a uniform way). The governing equation must be formulated on a different model. In section A of the appendix to this chapter, it is shown that the velocity of a wave in deep water is governed by the relationship

$$v = \sqrt{\frac{g\lambda}{2\pi}} \tag{4.5}$$

Equation 4.5 shows that the velocity of waves in deep water is independent of water depth. Since the wave-induced motion of the water particles occurs only near the surface, this is to be expected. The velocity, however, does depend on the wavelength, λ, whereas in shallow water it is the same for all waves regardless of wavelength. The longer the wave in deep water, the faster it travels. The motion of the water particles during passage of a deep water wave is quite different than in the case of shallow water. At a crest the particle velocity is horizontal and in the direction of advance of the wave (Fig. 4.11). At a trough it is again horizontal but in the opposite direction. At the places where $y = 0$ the particle motion is vertical. When the motion of an individual particle is traced out it follows, as the wave passes, a path which is circular (Fig. 4.12). Below the surface the size of the circle and, hence, the particle velocity becomes rapidly smaller with depth. At a depth equal to a half wavelength the motion due to passage of the waves has virtually ceased. A submarine need not submerge to a very great depth to avoid being tossed about by storm waves.

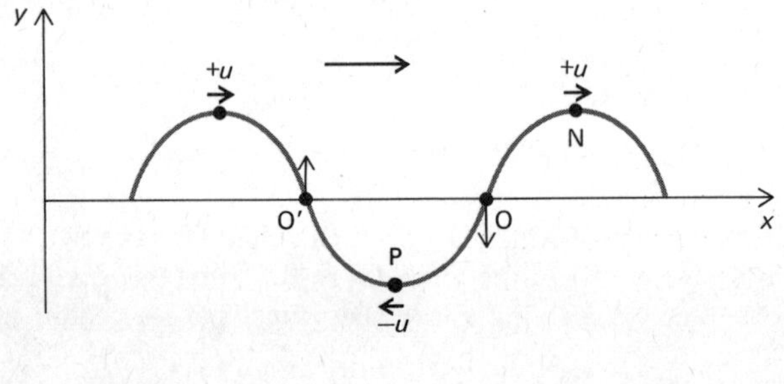

Figure 4.11 Direction of motion of particles on the surface of a deep water gravity wave. At points O and O′ the direction of the horizontal component of the velocity reverses.

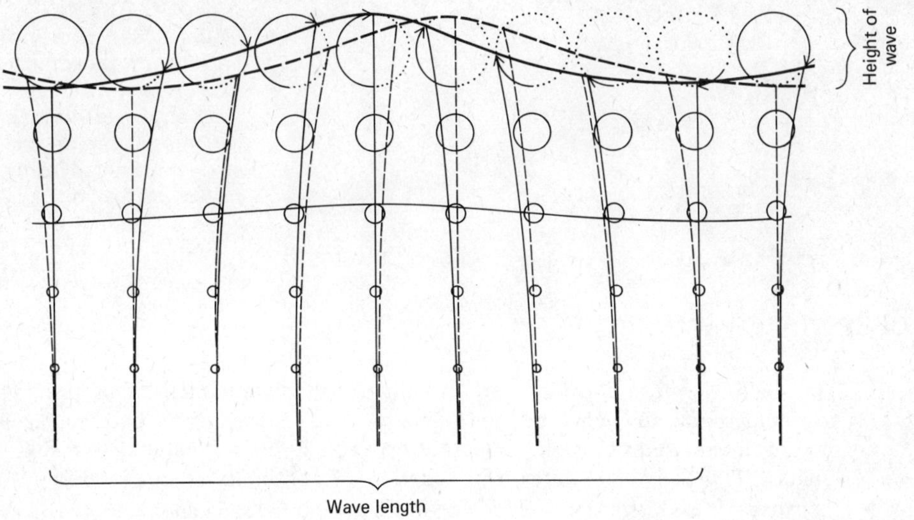

Figure 4.12 Motion of water particles during passage of a wave. The dashed line shows the corresponding positions when the wave has advanced a distance of $\lambda/10$. Individual particles trace out paths which are nearly circular. At greater depths the amplitude of the motion is smaller. From Zenkovich, *Processes of Coastal Development.* New York: Interscience; 1967.

Because the velocity, frequency, and wavelength of deep water gravity waves are all interrelated, each can be expressed in terms of the other. The relations are recorded in Table 4.1.

Table 4.1 Relations among Wavelength, Velocity, and Period for Deep Water Gravity Waves[a]

	λ	v	P
λ	λ	$\sqrt{g\lambda/2\pi}$	$\sqrt{2\pi\lambda/g}$
v	$2\pi v^2/g$	v	$2\pi v/g$
P	$gP^2/2\pi$	$gP/2\pi$	P

[a] *Note:* The quantity at each column head is expressed in terms of the quantity at the left. Thus, $v = gP/2\pi$.

If, for example, the period is observed, λ and v can be calculated with the corresponding relations in the bottom row of Table 4.1. A wave in deep water with a period of 22.5 sec, about the largest likely to be observed, has a wavelength of 792 meters and a velocity of 68 knots.

WAVES IN WATER OF INTERMEDIATE DEPTH

The velocity of waves on the surface of water has been calculated for the two special cases of deep and shallow water, with a quite different result being obtained in each case. As waves approach a gently shelving shore they must pass a region where

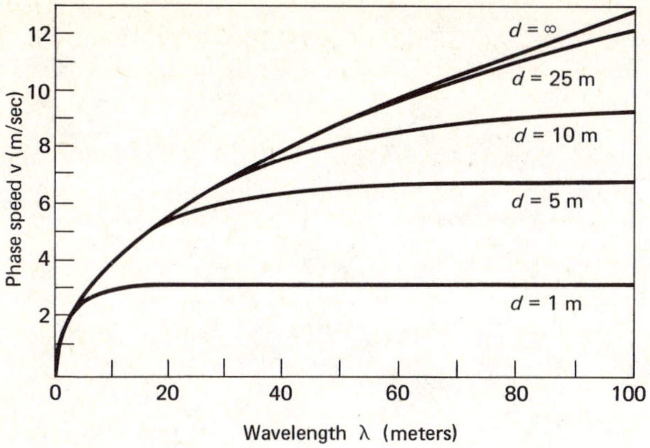

Figure 4.13 Velocity of waves as a function of wavelength in water of various depths. When the velocity curve is horizontal, velocity is independent of wavelength and the wave is in shallow water.

neither of these approximations holds very well. Thus it is important to know how each of the approximate solutions relates to the solution of the complete, or general, problem. This is shown mathematically in section B of the appendix to this chapter and graphically in Figure 4.13, which shows the velocities computed from a complete solution of the equations of wave motion. In deep water the speed of the waves increases steadily as the wavelength increases; in shallow water v soon becomes constant. Figure 4.13 shows that the transition from the deep to shallow water behavior is quite abrupt. This is why the two approximate cases analyzed above are adequate for dealing with the great majority of water wave problems.

CAPILLARY WAVES

Water has the property of *surface tension.* Consider a spherical drop of water. A sphere is the shape having the smallest surface area for a given volume. If the shape of the spherical drop of water is changed in any way, work must be done on it. This is the work required to create additional surface area. The work required to create a unit area of new surface is called the *surface energy.* Surface energy is equivalent to *surface tension,* the force per unit length required to create new surface. A beetle is able to walk on the surface of water because each of its feet distorts the water surface as in Figure 4.14. This distortion is resisted by the surface tension of the water which generates a force supporting the beetle. (Once the beetle's leg penetrates the surface, it is no longer supported and the beetle sinks.)

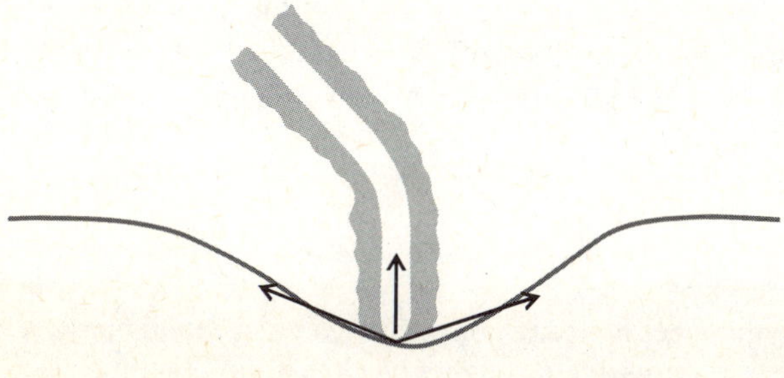

Figure 4.14 Enlarged view of foot of beetle walking on water. Surface tension forces which support the beetle are indicated by arrows.

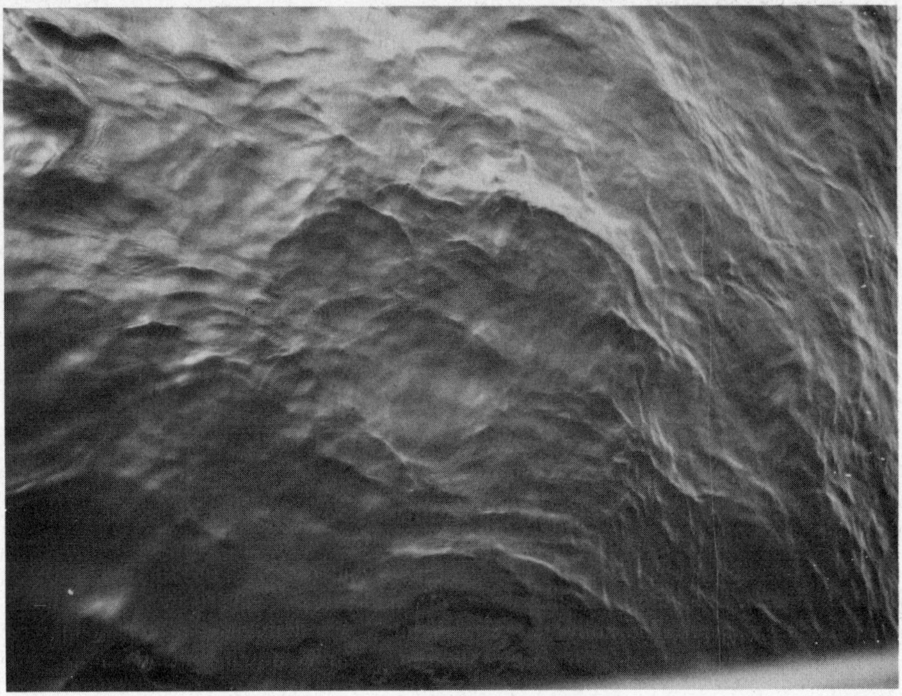

Figure 4.15 Capillary waves forming on the backs of larger waves.

Because surface tension resists the distortion of a water surface it clearly influences the formation of water waves. Only for very short waves is the surface tension force comparable to or greater than the gravitational force acting on the water in a wave. Very short waves, where surface tension forces dominate, are called *capillary waves* (because it is surface tension that makes water rise in a fine tube, a *capillary*). Theory shows that the velocity of capillary waves depends on wavelength, but in the opposite way from that of deep water gravity waves: For capillary waves the shorter the wavelength, the greater the velocity. Waves with very short wavelength travel with high velocity, and as the wavelength increases, the velocity decreases. In contrast, at large wavelengths, waves respond primarily to gravitational forces and their velocity increases as the wavelength increases. It follows, then, that there must be some critical wavelength for which the wave velocity is a minimum. No wave in deep water can travel with a slower speed. This critical wavelength is 1.73 cm and the minimum velocity is 23 cm/sec.

The wavelength of 1.73 cm is the upper bound of the domain of capillary waves. They are really only ripples on the surface, often seen on the backs of larger waves, (Fig. 4.15). The familiar catspaws produced by gusts of wind are examples of capillary waves.

GENERATION OF WAVES

Suppose that a storm commences at sea. Continued high winds build up wave amplitude and the sea becomes progressively rougher. As the wind continues the waves tend towards longer periods and greater wavelengths; the scale of a fully developed sea in the North Atlantic is impressive. There are two reasons for the increase in wave period: Waves of large wavelength can attain a greater height before breaking (which occurs when the angle at the wave crest is less than 120°), and they travel faster, more nearly at the wind speed.

A heavy sea is not generated instantaneously. The duration of the storm must be long enough and the *fetch,* the unobstructed distance over which the wind blows, must be sufficiently great. The role of these factors is illustrated in Table 4.2 in which

Table 4.2 Sea State Raised by Storms[a]

	Wind Speed (Beaufort Number)											
	3			*5*			*9*			*11*		
Fetch (Nautical miles)	*t*	*H*	*P*	*t*	*H*	*P*	*t*	*H*	*P*	*t*	*H*	*P*
10	4	0.6	2	3	1	3	2	2.5	4	2	3	5
40	12	0.6	3	9	2	4	6	5.5	6	5	7	7
160	43	0.6	5	27	2.5	6	18	10	9	16	13	10

[a] Data from Bowditch, *American Practical Navigator,* through courtesy U.S. Naval Oceanographic Office.

is recorded, for each fetch (in nautical miles) and wind speed (expressed as a Beaufort Number, see Appendix I), the minimum time, *t*, in hours that the wind must blow to raise waves of height *H* (meters) and period *P* (sec). A certain minimum fetch and duration is required for wind of a given speed to raise the waves it is capable of generating. A force 5 wind must blow over 160 miles of ocean for 27 hours to raise the full 4.2 meters of wave amplitude it can sustain. The table shows that the longer the fetch, the greater the period, and hence the wavelength, of waves. Waves of great height and long period are generated only by storms sweeping over the great open sea spaces (such as the Southern Ocean); in these places ships encounter the heaviest seas.

The waves generated by storm winds spread well out beyond the storm center. The storm-generated sea that runs in the absence of strong wind is called *swell.* Because wave velocity increases with wavelength (Eq. 4.5) the waves of longest wavelength generated in the storm travel most rapidly outwards from the center; waves of shorter wavelength travel more slowly and fall behind the long wavelength waves. At sea the first warning of an approaching storm is the arrival of a marked swell of long period. As the storm approaches, waves of shorter period are encountered while in the storm center a ship finds herself surrounded by a confused sea where waves of many periods and wavelengths are represented.

An estimate of the distance to the source of a swell can be made by observing how rapidly the period of the waves reaching the shore decreases. The more distant the source, the greater will be the separation of the short and long period waves (since the velocity difference has been acting for a longer time). From analysis of wave spectra, swell arriving on the coast of Cornwall, England, has been traced to a storm in the vicinity of the Falkland Islands, the waves having traversed almost the whole length of the Atlantic Ocean. (The speed with which wave energy travels outwards from a storm center is calculated in appendix C to this chapter.)

Often the swell arriving at a beach is composed of contributions from two or more storm centers. Usually the swell from each source will have a different period; the

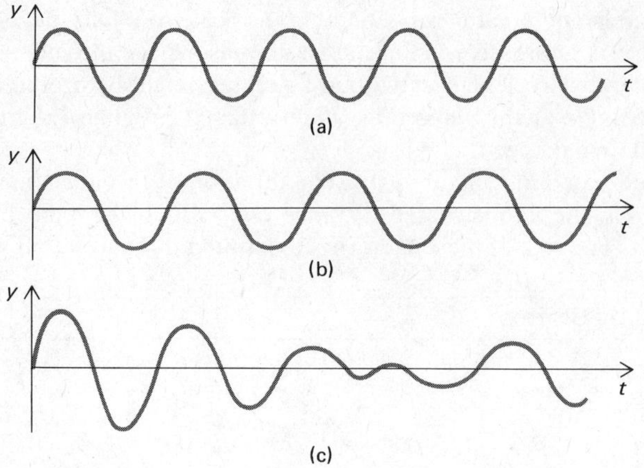

Figure 4.16 Surf beat can arise from the arrival of swell from two different sources. Curve (c) is the sum of (a) and (b).

stronger and more distant the storm, the longer the period. The waves arriving at a beach may be the sum of two wave trains of different period and amplitude as in Figure 4.16a and b. The amplitude of the resultant waves, shown in Figure 4.16c, will then vary in a regular way. The *beat frequency* is the frequency with which the minimum wave amplitude of Figure 4.16c occurs. An experienced surfboat coxswain will always study the surf for a time before attempting a launch. He determines the beat frequency in units of the wave period. He then can judge the best time to launch his boat (just before the minimum in the beat) and the length of time that his crew will have to pull through the line of breakers before high amplitude waves arrive again. The difference between a successful rescue and a swamped surfboat lies in astute observation of the sea. Through his ability to make really careful, scientific observations, the master mariner is able to save lives when the less skilled observer loses all.

SURF

Waves in deep water present difficulties to quantitative description, but these are essentially the mathematical problems encountered in solving the equations that describe the water motion. The physics of such waves is relatively simple. The description of waves as they approach shore and become surf presents a different type of difficulty: Even identification of all the major physical processes occurring is difficult. As a wave approaches shore and breaks, the water motion becomes turbulent and highly complex; energy dissipation becomes large. Further, waves near shore often interact with the bottom, placing quantities of sand or mud into suspension. These processes also strongly absorb wave energy and so complicate a quantitative description of the wave motion. Finally, in shallow water the wave velocity depends on water depth. Each shore has its own characteristic slope, perhaps with one or more offshore bars present. The wave pattern on every slope will be different. Here only some of the more important near-shore processes can be identified and described; the observer must decide on their relative importances for whatever shore happens to interest him.

In order to discuss the formation of surf, consider the approach of waves to a gradually shoaling beach. As the water becomes shallower, the waves begin a transition from the deep water to the shallow water type. Where the depth equals the

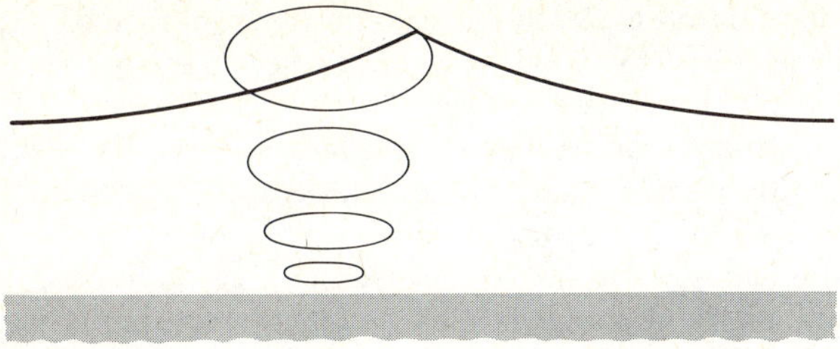

Figure 4.17 As waves approach shore the orbits traced out by the water particles are flattened. At the bottom only horizontal motion occurs.

wavelength, the bottom has scarcely any effect on the wave at all. When the depth is a third of the wavelength, the wave begins to slow down appreciably and the motion of the water particles at the surface begins to depart from a circular orbit. Below the surface the particle orbits are flattened even more until, as illustrated in Figure 4.17, the particles near the bottom move horizontally.

As a wave moves into shallow water its shape changes in a characteristic way. When the wave height begins to become comparable to the depth, the wave crests are moving in substantially deeper water than the troughs. Each crest moves with a higher speed and tends to overtake the slower-moving trough ahead of it. The leading edge of the wave is consequently steepened. The orbital velocity of a water particle at the crest is high but in the trough it is reduced by bottom friction. Both these factors tend to make the wave unstable. When the wave height becomes about 1.3 times the mean water depth there is no longer enough water ahead of the crest to fill it out in its advance. The crest becomes unsupported and the water at the crest spills down the front of the wave in a foaming, turbulent cascade. The wave has become a breaker. The exact form taken by the breaker depends on the slope of the shore relative to the wave size. If the shoaling is gradual and uniform, spilling breakers form (Fig. 4.18). Breaking occurs over a considerable time during which the wave retains

Figure 4.18 Two spilling breakers approaching shore. In the foreground the turbulent water of the bore formed by the preceding breaker advances up the beach; the water here appears lighter in color because of suspended sand.

its organization. A surfboard is most effectively used on the front face of a spilling breaker and the best beaches for such activity are those where there is a small tidal range and a long, gentle slope out to deep water.

When shoaling is abrupt relative to the wave size, plunging breakers occur (Fig. 4.19). Here the top of the wave remains sharp as it overturns and drops down in front of the advancing wave. Often air is entrapped and compressed by the collapsing wave. In a large wave this compressed air is released in an eruption of foam and spray (Fig. 4.20). The duration of the plunge is short and the wave loses its organization, becoming a mass of turbulent water traveling shoreward as a bore over the return current from the previous wave. Often, a number of bores will be seen, those further out advancing on those further in, because they are in deeper water and travel faster. In the acts of breaking and bore formation the energy of the wave (which may have been acquired in a storm thousands of miles away) is suddenly dissipated.

It is only inside the line of breakers, in the zone of advance of the bore up the beach slope (the *swash* zone, the foreground of Figure 4.18) that any significant return current exists. The return current may be swift, sufficient to knock a person off his feet, but it is always in relatively shallow water and does not extend seawards beyond the line of breakers. The "undertow" of which swimmers are so often warned simply does not exist; the net transport of water by waves up until they are destroyed by breaking is very small. Localized rip currents, to be discussed in the next chapter, may run seaward, but there is no general seaward current due to wave action to carry a swimmer away from the beach.

Figure 4.19 Development of a plunging breaker. Because the wave is approaching at a slight angle to the shore it is breaking progressively and each stage of the breaking process can be seen in the photograph.

Figure 4.20 Waves striking Minots Lodge off the Massachusetts coast often throw spray to the lighthouse lantern 85 ft. above sea level. (Rapho Guillumette Pictures).

REFLECTION OF WAVES

When a coastline is formed by a steep cliff rising from deep water, waves having wavelengths less than the water depth have no chance to break before striking the cliff. At the cliff face only vertical motion of the water particles is possible. As a result, the incoming wave is reflected from the cliff and returned seawards. The reflection of sea waves follows the same rule as the reflection of light waves: If the arriving waves are parallel to the reflection surface, they are reflected with only a reversal of direction; if they are inclined, the reflected wave is inclined as shown in Figure 4.21. Incident and reflected waves interfere with each other. Where their crests pass through each other, as at A in Figure 4.21, the wave peaks up to great steepness without breaking. When the incident and reflected waves are parallel, a standing wave pattern may be generated.

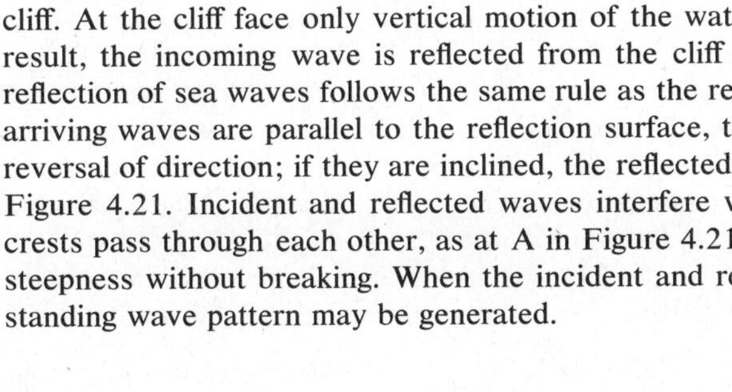

Figure 4.21 A wave crest (solid line) is reflected at the sea wall at the inclination shown by the dashed line. Maximum amplitude occurs when an incoming and a reflected wave cross at A.

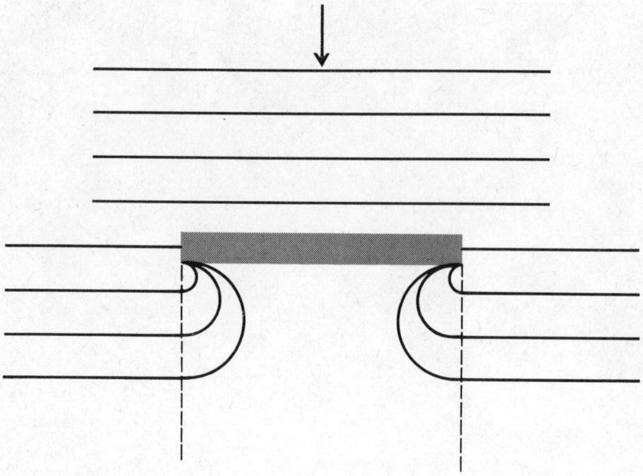

Figure 4.22 The wave shadow cast by the breakwater is not sharp because of diffraction around its ends.

When a wave approaching the shore is reflected, its energy is returned to the sea surface. It follows that the most effective way of protecting a shore from waves is to build a breakwater offshore in water so deep that no breakers occur; a "wave shadow" has been created. Ideally, the breakwater should have a smooth, vertical face to reflect the incoming waves. In practice it is usually difficult to meet these conditions; for reasons of economy, breakwaters are often built in too shoal a depth and with rough and sloping faces. Only partial reflection is achieved and the breakwater structure must be heavy enough to dissipate the unreflected wave energy without structural damage. In exposed locations this may require a massive structure. For example, the outer 50 meters of the heavy stone breakwater that forms the harbor at Wick (Scotland) has been detached and carried inside the harbor by severe storm waves.

Another factor which limits the effectiveness of an offshore breakwater is the diffraction of waves around its ends. If the waves were of very small wavelength, the breakwater shown in Figure 4.22 would cast the shadow indicated by dashed lines; between these lines the water would be calm. In fact, some wave energy leaks around the corners of the breakwater, the phenomenon known as *diffraction*. It arises because a wavefront at any given instant of time may be thought of as a source of new waves which spread out in all directions. When the wavefront is straight and continuous, interference eliminates all the new waves except those directly ahead; the wavefront advances in a straight line. At a corner, at the end of the breakwater, the new waves fan outwards without hindrance. These are the diffracted waves which move into the area behind the breakwater. An analogous phenomenon occurs when a beam of light passes a sharp straight edge or through a pin-hole or slit, giving rise to optical diffraction patterns.

REFRACTION

If straight, parallel waves advance directly up a uniformly sloping beach, the waves remain straight and parallel but their spacing decreases as they slow down in the shallower water. This is illustrated in Figure 4.23a. Suppose that the waves are not headed straight at the beach but are approaching at an angle as in Figure 4.23b. So

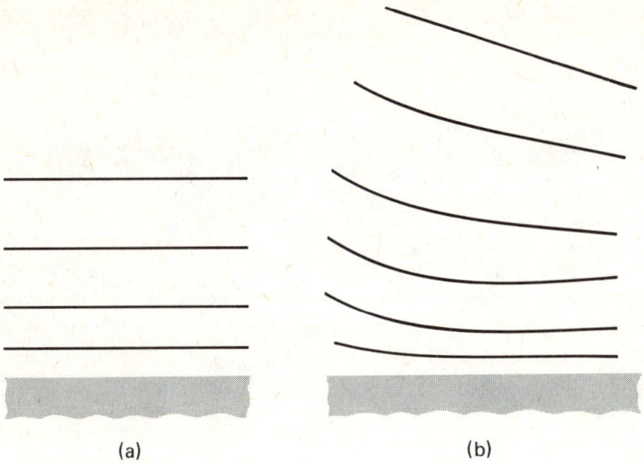

Figure 4.23 Waves advancing straight up a beach remain parallel to the shore (a) while waves approaching at an angle are refracted (b) so as to approach the shore more nearly straight-on.

long as the waves are in deep water they remain parallel. As they enter shallow water they are slowed. The inshore end of any given wave will be slowed more (because it is in shallower water) than the end further offshore. It follows that the wavefronts will become bent, as in Figure 4.23b. This is called *refraction*. (The corresponding optical phenomenon is the bending of light rays, lines everywhere perpendicular to the wavefronts, as the light passes through media causing different amounts of refraction). As a result of refraction, waves always tend to approach a beach parallel to the shore (Fig. 4.24).

An important consequence of wave refraction is the focusing of wave energy onto promontories, as illustrated in Figure 4.25. Wave refraction is, consequently, a factor tending to smooth out an irregular shoreline. The promontories are eroded more rapidly because of the focusing on them of wave energy. Waves are a major factor shaping the coast, a subject which is explored in the next chapter.

APPENDIX

A. Velocity of a deep water wave.

A simple calculation of the velocity of a wave in deep water can be made by using the kinetic energy of a particle of water moving over the wave surface, provided the form of the wave is a simple, symmetrical one. Consider, then, a sinusoidal wave train of small amplitude (A). At a particular instant of time the water surface is represented by

$$y = A \sin \frac{2\pi}{\lambda} x$$

This equation is plotted in Figure 4.26. It has the property that whenever x increases by λ, the surface pattern is repeated [$\sin (2\pi/\lambda \cdot \lambda = \sin 2\pi = \sin 0 = 0$].

The direction of motion of surface particles is shown in Figure 4.11. This particle motion makes the crests and troughs advance. At points O and O′ in Figure 4.11 the direction of the horizontal component of the particle velocity reverses. Now, in order to facilitate the study of a particular wave, suppose that the water across which the

Figure 4.24 Waves advancing across this bay from left to right are refracted as they pass the small promontory so as to swing in towards the beach in the foreground.

waves are advancing with velocity v is caused to flow in the opposite direction with velocity $-v$. The wave patterns will then appear to stand still and a surface particle will flow over successive crests and troughs from right to left. The horizontal components of the particle velocities at points N, O, and P are easily derived from Figure 4.11; they are shown in Figure 4.27. At O the horizontal particle velocity is equal to the magnitude of the wave velocity. This is the quantity to be calculated.

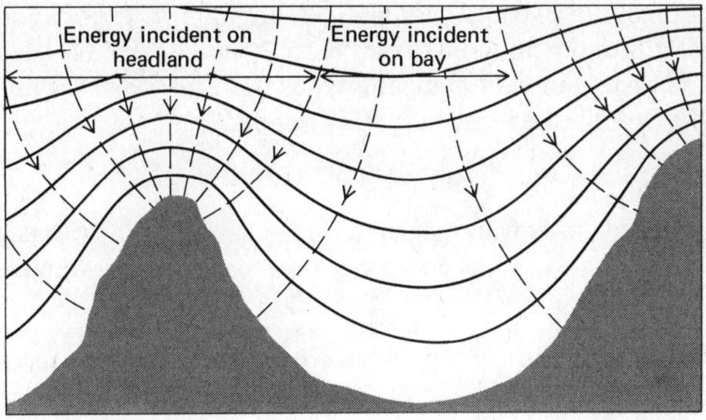

Figure 4.25 Waves approaching an irregular shore are focused on the promontories by refraction. This directs a relatively large amount of energy onto the promontories as compared to the embayments. The resultant erosion tends to straighten the coast line.

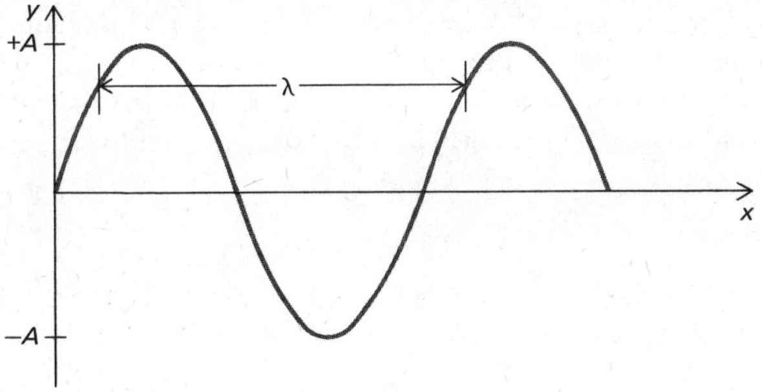

Figure 4.26 Graph of the equation $y = A \sin(2\pi x/\lambda)$. The distance λ is the wavelength. Points separated by λ are at exactly the same elevation and have the same slope.

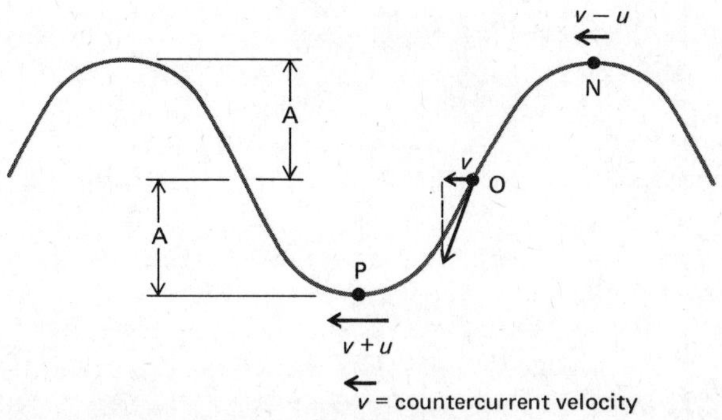

Figure 4.27 Particle motions in the wave of Fig. 4.11 when a countercurrent velocity, v, is applied to the water so as to hold the wave pattern stationary.

First, consider the passage of a particle from point N to point P. At N and P the velocity is in a horizontal direction; in going from N to P the particle has dropped down the back side of the crest, gaining kinetic energy as it falls. By conservation of energy this must equal the potential energy lost. In symbols,

$$2mgA = \tfrac{1}{2}m(v+u)^2 - \tfrac{1}{2}m(v-u)^2 \qquad (4.6)$$

This follows because the force on a water particle of mass m is mg, it has fallen through a distance $2A$, and lost potential energy $2mgA$. Expand the squares and collect terms to find

$$2mgA = \tfrac{1}{2}m(v^2 + 2uv + u^2 - v^2 + 2uv - u^2) = 2muv$$

or

$$gA = uv \qquad (4.7)$$

Note that in making this energy balance we are assuming that the water and the air above it have no viscosity so that there is no friction encountered by the particle sliding over the surface. This turns out to be a good approximation for ocean waves. Note also that the only force acting to restore a water particle displaced by the wave to its position of rest is gravity. These waves are therefore called *gravity waves.*

Equation 4.7 contains two unknowns. To find v, another condition must be imposed on the particle motion, a condition which expresses the fact that the waveform is symmetrical. This is obtained by examining the velocity of a surface water particle at point O. As shown in Figure 4.28 the velocity at O is of magnitude s and is in the direction of a vector drawn tangent to the surface. The horizontal and vertical components of s are v and w. Hence

$$s^2 = v^2 + w^2 \qquad (4.8)$$

The kinetic energy of the water particle at O is the kinetic energy it had at the top of the wave plus the additional kinetic energy acquired in falling halfway down the back side of the wave. Thus

$$\tfrac{1}{2}ms^2 = \tfrac{1}{2}m(v-u)^2 + mgA \qquad (4.9)$$

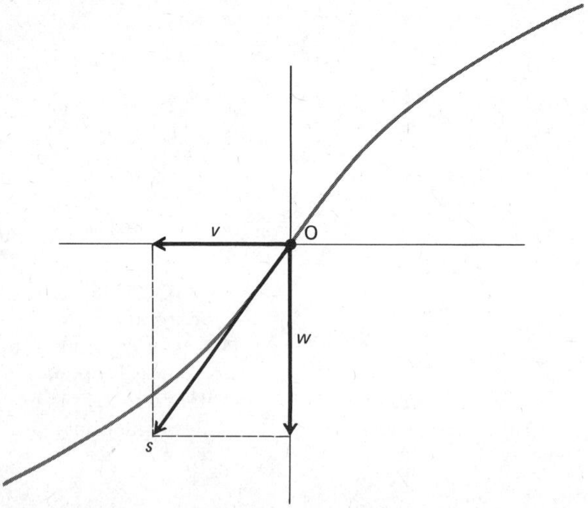

Figure 4.28 Particle velocity vectors at the point O.

Solve Eq. 4.9 for mgA and substitute for s from Eq. 4.8 to find

$$\begin{aligned} mgA &= \tfrac{1}{2}m(v^2 + w^2) - \tfrac{1}{2}m(v-u)^2 \\ &= \tfrac{1}{2}m(v^2 + w^2 - v^2 + 2uv - u^2) \\ &= \tfrac{1}{2}m(w^2 - u^2 + 2uv) \end{aligned} \qquad (4.10)$$

Now Eq. 4.7 shows that $gA = uv$. Hence we can eliminate uv from 4.10 to get

$$mgA = \tfrac{1}{2}m(w^2 - u^2 + 2gA)$$

or

$$w^2 = u^2 \qquad (4.11)$$

Thus, in a symmetrical wave, the vertical component of the particle velocity at O is equal in magnitude to the horizontal components at N and P when the hypothetical countercurrent is turned off.

Now make use of the condition that the wave is of low amplitude and of sinusoidal form. At a given instant of time the displacement of the surface of the water is

$$y = A \sin \frac{2\pi}{\lambda} x \qquad (4.12)$$

In a small interval of time, δt, the displacement of the particle at O will be $s\delta t$ which has displacement components $\delta y = w\delta t$ and $\delta x = v\delta t$. Since the motion of the particle must follow the water surface, Eq. 4.12 requires that

$$w\delta t = A \sin \frac{2\pi}{\lambda} v\delta t$$

Since δt can be made as small as desired, the quantity $(2\pi/\lambda)v\delta t$ can be made to be a small quantity. Since for a small z, $\sin z \simeq z$, and since from Eq. 4.11 $w = u$, it follows that

$$w\delta t = u\delta t = A \frac{2\pi}{\lambda} v\delta t$$

or

$$u = \frac{2\pi}{\lambda} Av$$

Equation 4.7 shows that $gA = uv$. Hence

$$u = \frac{gA}{v} = \frac{2\pi}{\lambda} Av$$

or

$$v^2 = \frac{\lambda g}{2\pi}$$

The velocity with which the crests and troughs in deep water advance is then

$$v = \sqrt{\frac{g\lambda}{2\pi}} \qquad (4.5)$$

B. Velocity of waves in water of any depth

As indicated previously, an exact solution to the hydrodynamic equations cannot be obtained for gravity waves on water. However, solutions to any desired degree of accuracy can be calculated by a procedure known as perturbation theory. The first approximate solution for the velocity is

$$v = \sqrt{\frac{g\lambda}{2\pi} \tanh \frac{2\pi d}{\lambda}} \tag{4.13}$$

The results obtained by simple physical arguments as Eqs. 4.4 and 4.5 are special cases of Eq. 4.13. This is a consequence of the property of the hyperbolic tangent function (tanh) that $\tanh z \simeq 1$ for $z \gg 1$ and $\tanh z = z$ for $z \ll 1$. In deep water $d \gg \lambda$, so that $2\pi d/\lambda \gg 1$ and $\tanh (2\pi d/\lambda) \simeq 1$ and Eq. 4.13 reduces to Eq. 4.5. In shallow water, $d \ll \lambda$, $2\pi d/\lambda \ll 1$ and $\tanh (2\pi d/\pi) \simeq 2\pi d/\lambda$. Then $v \simeq \sqrt{gd}$, which is Eq. 4.4. The curves in Fig. 4.15 are calculated from Eq. 4.13.

C. Group velocity

The waves raised by the wind at sea are not simple trains with uniform wavelength, but the complex surface of the actual sea may be thought of as resulting from the interference of several wave trains of different wave lengths. If these trains of waves are traveling in shallow water the interference pattern will be stable because all the wave components travel with the same velocity. Ordinarily, this would occur only very near the shore. The waves raised at sea are deep water waves, for which the velocity increases as the wave length increases. In a complex sea, then, the long waves will advance ahead of the short ones and the form of the interference pattern will continually change. As a result, the place where the wave amplitude is largest will move ahead at a velocity (the *group velocity*) different from that of any individual wave (the *phase velocity*).

To make a quantitative determination of the group velocity, consider the advance of two wave trains having nearly equal wavelengths, λ and $\lambda + \epsilon$. These are illustrated in Figure 4.29. At point A waves 1 and 2 are exactly in phase; constructive

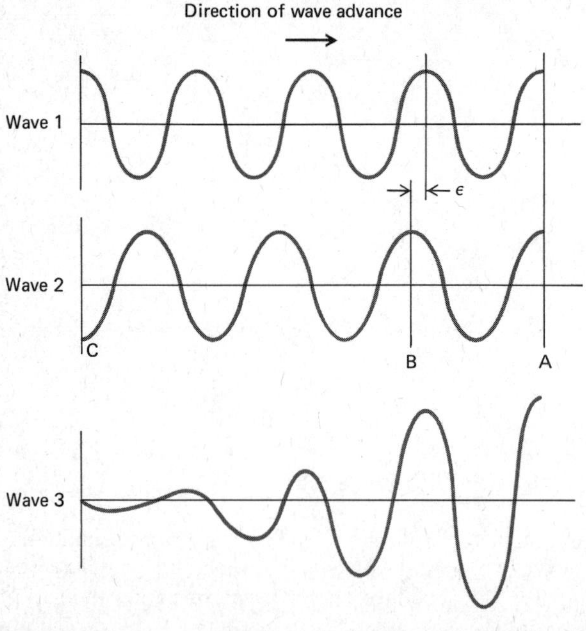

Figure 4.29 A wavetrain, 3, consisting of two components, 1 and 2, whose wavelengths differ by ϵ.

interference occurs and the resultant wave is of largest amplitude. At point C waves 1 and 2 are exactly out of phase and the resultant amplitude is zero. Wavetrain 1 travels with velocity V; 2, with velocity $V + v$. Hence the longer waves overtake the shorter with velocity v. Wavetrain 2 will pick up distance ϵ on wavetrain 1 in time $t = \epsilon/v$; in this time the position of greatest total amplitude will fall back from A to B, a distance of one wavelength. The velocity at which it falls back is

$$w = \frac{\lambda}{t} = \frac{\lambda v}{\epsilon} \tag{4.14}$$

This equation applies for any kind of dispersive wave (a wave where the velocity depends on wavelength). Now, consider deep water waves explicitly. For these waves

$$V = \sqrt{\frac{\lambda g}{2\pi}}$$

$$V + v = \sqrt{\frac{(\lambda + \epsilon)g}{2\pi}}$$

In order to find v/ϵ take the difference of the square of the velocity,

$$(V + v)^2 - V^2 = \frac{(\lambda + \epsilon)g}{2\pi} - \frac{\lambda g}{2\pi} = \frac{\epsilon g}{2\pi}$$

$$V^2 + 2Vv + v^2 - V^2 = \frac{\epsilon g}{2\pi}$$

$$2Vv + v^2 = \frac{\epsilon g}{2\pi}$$

$$(2V + v)v = \frac{\epsilon g}{2\pi}$$

Because $\epsilon < \lambda$, $v \ll V$ so that $2V + v \simeq 2V$. Making this approximation,

$$2Vv = \frac{\epsilon g}{2\pi} \tag{4.15}$$

or

$$\frac{v}{\epsilon} = \frac{g}{4\pi V}$$

This result may be substituted in Eq. 4.14 to find that

$$w = \lambda \frac{v}{\epsilon} = \frac{g\lambda}{4\pi V} = \frac{V^2}{2V} = \frac{V}{2}$$

since $\lambda g/2\pi = V^2$. The speed of advance of the place of maximum amplitude, the group velocity, is then

$$V_g = V - w = V - \frac{V}{2} = \frac{V}{2}$$

When waves are generated by a storm at sea, components of varying wavelengths are created and spread outwards from the storm center. Because of interference between the various wave components, the resultant waves of greatest amplitude advance at speed V_g, half the speed at which a simple wavetrain would travel. The wave energy is propagated at speed V_g. In calculating the time required for waves to appear on the beach at Nantucket because of a storm centered at Bermuda, $V_g = V/2$, calculated for the average wavelength, provides the answer.

Further Readings

BARBER, N. F., *Water Waves.* London: Wykeham, 1969.

BASCOM, W., *Waves and Beaches.* New York: Anchor Books, 1964.

NEUMANN, G., and PIERSON, W. J., JR., *Principles of Physical Oceanography.* Englewood Cliffs, N.J.: Prentice-Hall, 1966 (Chapters 10 and 12).

TRICKER, R. A. R., *Bores, Breakers, Waves and Wakes.* New York: Elsevier, 1964.

5 Interfaces II: Wind and Water on the Land

All of the face of the solid earth is exposed to the wearing action of wind and water. The rate at which the hard rock of the continents is worn away is fast enough to cut North America down to sea level in 14 million years. Since central North America has been exposed to the elements for over a thousand million years, this is a relatively short time. (Chapter 8 explains why there are still mountains in North America.) In places where the surfaces are not hard rock but are covered with sand, mud or soil, the wearing action of wind and water is often fast enough to observe in a period of minutes or hours. Erosion and deposition of soil and sand is of immediate practical importance to the use and management of land in agriculture, in the development of water resources, and in coastal engineering. This chapter explains the physics of the movement of land by wind and water.

FLUID FLOW

The force which a moving fluid exerts on a solid surface depends on the character of the flow, the velocity, and the boundary between fluid and solid. The simplest type of flow is *steady;* there are no changes of the flow pattern with time. The pattern in which the fluid is flowing can be visualized by looking at its *streamlines,* imaginary lines drawn through the fluid so as to be always parallel to the direction of fluid motion. Streamlines can be traced out in laboratory experiments by injecting a small amount of dye (smoke if the fluid is air) into the fluid by a fine jet held at a fixed position, as in Figure 5.1. In steady flow the pattern of streamlines is fixed. Where the streamlines converge, as at A in Figure 5.1, the velocity of the fluid increases. At A the size of the pipe is smaller; the fluid has to move faster if there is not to be an accumulation of fluid upstream. Where the streamlines diverge, the velocity decreases.

The flow illustrated in Figure 5.1 can also be described as *laminar;* the streamlines at the edges of the fluid are parallel to the tube walls. Laminar flow is relatively easy to describe mathematically; unfortunately most fluid flow in nature is not laminar but turbulent. Laminar and turbulent flow are illustrated in the experiment (first performed by Osborne Reynolds) shown in Figure 5.2. Fluid is flowing through the tube and marker dye or smoke is injected. In Figure 5.2a the velocity is low and the flow is

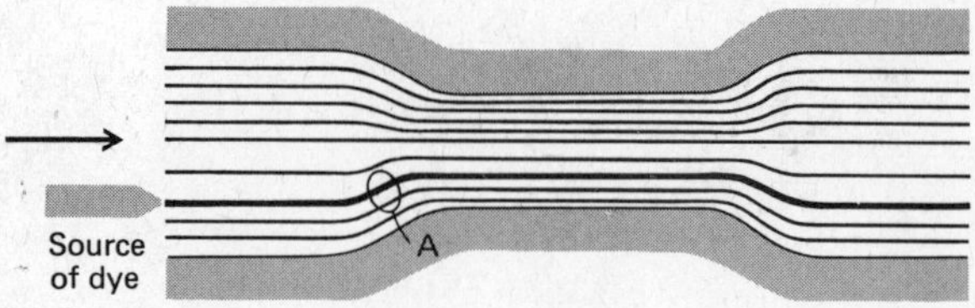

Figure 5.1 A streamline is traced by dye injected from the nozzle into the liquid flowing through the pipe. Additional streamlines for this flow pattern are drawn in. At A the streamlines draw together as the fluid velocity increases.

laminar throughout. When the flow velocity is increased the flow pattern suddenly changes from laminar to turbulent, as shown in Figure 5.2b. The turbulent pattern of fluid motion revealed by the thread of dye continually changes. The fluid velocity at any given point fluctuates in magnitude and direction (while maintaining a net flow downstream). Eventually the dye or smoke is thoroughly mixed throughout the fluid. Gusts on a windy day are due to turbulence in the flow of air over the ground. Turbulent flow in the main channel of the Delaware River can be seen in the foreground of Figure 5.3.

When the flow of wind or water becomes turbulent, the forces on the land below are greatly increased. Three factors determine whether fluid flow in a given situation will be laminar or turbulent: the velocity, the size of the channel in which the fluid moves, and the viscosity of the fluid. *Viscosity* is the resistance of a fluid to change in shape. A perfect fluid has no viscosity. Examples of fluids with high viscosity are glycerine, heavy engine oil, and, an extreme case, pitch. The higher the viscosity of a fluid, the more difficult it is to make its flow turbulent because viscous forces dampen the locally high velocities in the eddies. On the other hand, high velocity and large channel size favor development of turbulence.

FLUID DRAG

Suppose that there is a laminar fluid flow over a flat surface. The layer of fluid immediately adjacent to the solid surface does not flow because it adheres to the solid. Moving away from the surface, the fluid velocity steadily increases, as in Figure 5.4; at a distance δ above the surface it reaches the steady velocity characteristic of the main body of fluid, v. The layer of thickness δ over which the velocity changes from

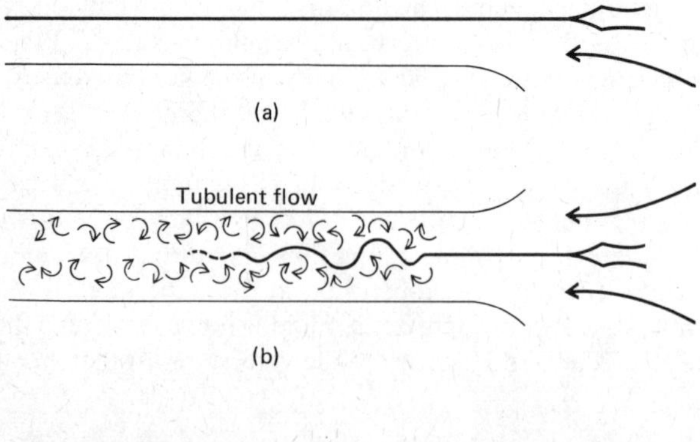

Figure 5.2 When flow through the tube is laminar, as in (a), the dye marker traces out a straight streamline. In (b) the flow velocity has been increased to the point where turbulence has started to develop.

Figure 5.3 The Delaware River looking north towards the Delaware Water Gap. The river channel is filled by the spring runoff and the flow is highly turbulent as can be seen in the foreground.

zero to v is called a *boundary layer*. The fluid above tends to drag the fluid in the boundary layer along with it. This drag force is transmitted through the boundary layer to the solid surface below. The fluid motion in the boundary layer may be turbulent rather than laminar if v is large, if the length of the surface in the direction of flow is great, or if the viscosity is low. It is often easy to see from on deck the turbulent boundary layer next to the hull of a ship which is under way. When the boundary layer is turbulent, drag is greatly increased.

Drag limits the speed with which a fluid can be pumped through a pipe. The drag on the hull of a ship or aircraft resists its forward motion and must be overcome by the thrust developed by the propulsion machinery if speed is to be maintained. The drag on an object moving through a fluid can be conveniently studied in a *free fall*

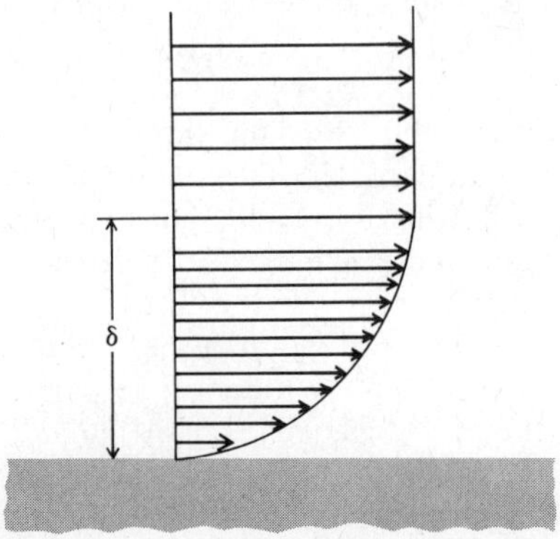

Figure 5.4 Laminar flow of fluid over a flat surface. The arrows represent velocity. Adjacent to the surface the fluid is at rest. Throughout the boundary layer of thickness δ the velocity increases; above the boundary layer it is constant. Because of the viscosity of the fluid, a drag force acts to the right on the flat surface.

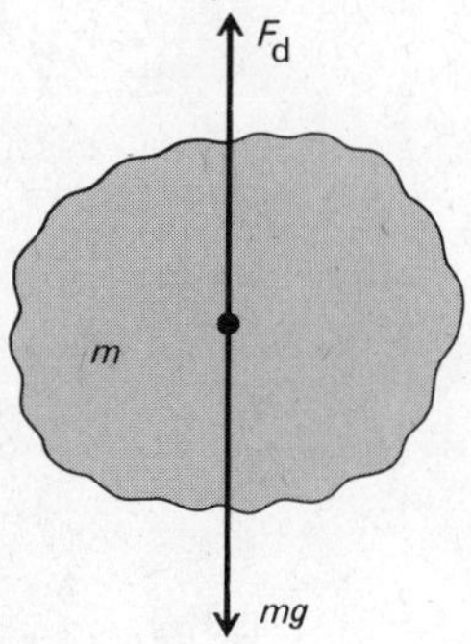

Figure 5.5 A grain of sand of mass *m* falling through a fluid is subject to the downwards acting gravitational force *mg* and the upwards acting drag force F_d.

experiment (Fig. 5.5). The object, say a grain of sand, is released at the top of a column of fluid. It is subject to a constant gravitational force of magnitude *mg* acting downwards. (The mass of the object is *m*; the acceleration due to gravity, *g*.) As the particle accelerates under this force, the upwards acting drag force increases. When these forces exactly balance, the particle continues to descend at a constant speed called the *terminal velocity*. An object dropped from an aircraft at high altitude quickly reaches its terminal velocity; it maintains this speed until it hits the ground. By attaching a parachute to increase the drag, the terminal velocity can be reduced so that a relatively soft landing can be made.

It is a common observation that fine sediment settles out of still water much more slowly than coarser material. It is easy to understand why this is so. The gravitational force is proportional to the mass of a particle, but the drag force (for particles of the same shape) depends on surface area. Taking a sphere of diameter *d*, for example, the volume is proportional to d^3 while the surface area is proportional to d^2. As *d* is decreased, the volume becomes smaller much faster than the surface. A fine particle has a relatively large surface area in proportion to its mass. In a free fall experiment, then, the drag on a fine particle is large relative to the gravitational force; its terminal velocity is small.

The same considerations govern the plucking of particles off a bed over which fluid is flowing. Suppose that the bed is composed of equal size particles between which there is no cohesion. Beach sand is a good approximation to this. A section of bed surface is shown in Figure 5.6. The grain at A is subject to a drag force, F_d, from the moving fluid, and to the gravitational force, *mg*, which tends to hold it in place. At the *threshold velocity* the drag force is just great enough to move the particle from its place on the bed. At any greater fluid velocity, particles will be continuously removed from the bed; that is, erosion will take place.

One would expect that the finer the particle, the lower the threshold velocity, since the drag force is relatively greater on small particles. Threshold velocity data are shown in Figure 5.7 for the case of water flowing over sand. The expected relationship holds for particles greater than 0.5 mm in diameter; higher current speed is required to move larger particles. But for smaller particles the threshold velocity increases with decreasing particle size. This is because we have been considering individual particles, whereas in a bed of fine sand grains the particles are closely packed. Below about 0.5 mm in diameter, they no longer respond as individuals to flowing water. A layer of nearly stagnant fluid clings to the particles at the bed surface; the drag force is spread over all the particles instead of acting on only the more

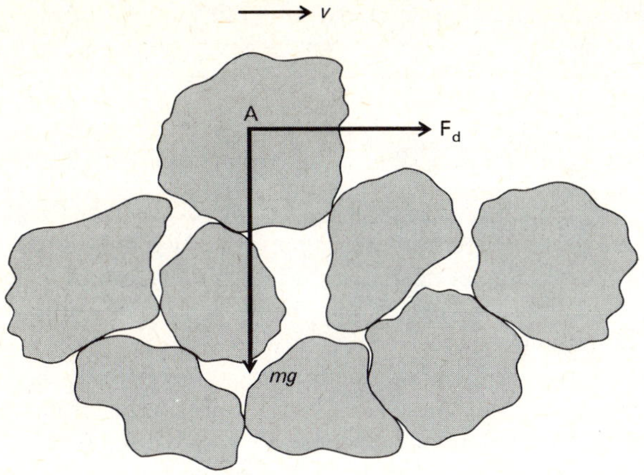

Figure 5.6 Fluid is flowing to the right over a bed of cohesionless sand grains. When F_d becomes large enough grain A will be picked off the bed by the fluid flow.

exposed grains. Hence, it is more difficult to set the very fine grain particles in motion. This effect may also be observed in an open field: A much higher wind velocity is required to raise dust from the ground composed of uniform fine particles than from a bed of coarser soil.

When air or water flows at erosional velocities over a bed of sand or clay the flow is turbulent. The eddies in the turbulent flow create local, upwards moving currents of fluid. The upward velocity exceeds the terminal velocity of the particles. Consequently, the particles are *suspended* and are carried along with the fluid. If the fluid velocity drops so that the terminal velocity is no longer exceeded in the updrafts, the suspended material will be deposited. The velocity curve below which deposition occurs is shown in Figure 5.7. Between the erosion and deposit curves in the graph there is a range of velocities which are insufficient to remove particles from the bed but are sufficient to keep in suspension particles already picked up. This is called the "transport" zone.

If a fluid carrying particles of various sizes in suspension is gradually slowed, the

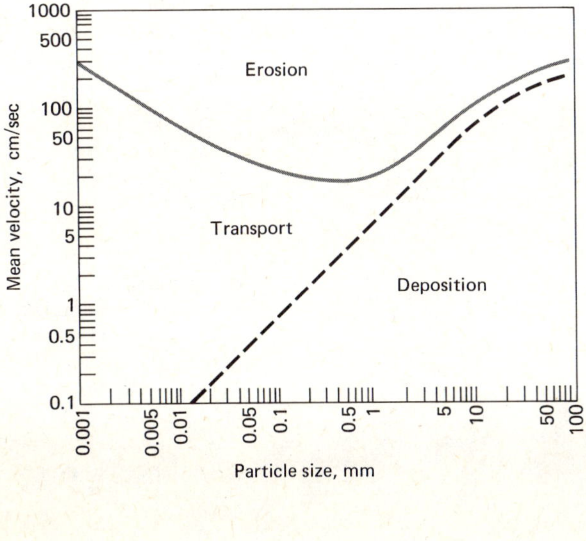

Figure 5.7 Mean velocity required to produce threshold drag on particles of various sizes is shown by the solid curve. The dashed curve shows the terminal velocity of fall in still water. The figure is constructed for particles having a density equal to that of sand grains. After Hjulstrom. Bul. Geol. Inst. Univ. Upsala, **25** 221 (1935).

coarser particles will be deposited first. For example, when turbid water from a fast-moving stream flows into a quiet lake, the sedimentary deposit on the lake bottom is progressively finer in texture outwards from the mouth of the stream.

WIND-BLOWN SAND

A sand desert is a good place to study the transporting power of wind. In a true sand desert the only material is sand grains of nearly uniform size. Fine particles which would become dust—suspended matter—are absent. There is no cohesion between the grains so the idealized conditions assumed in the introductory discussion are well approximated. When wind blows over a sand surface, grains set in motion by the wind collide with other grains. In most such collisions one or the other of the colliding grains is knocked high into the air. Because of wind drag, such a grain acquires considerable forward velocity as it falls back to the ground. A characteristic path is shown in Figure 5.8. When the falling grain strikes the ground another grain is thrown up into the air to continue the process. This transport of solid grains is called *saltation*. It accounts, typically, for 75 percent of the volume flow of wind-driven sand. The remaining 25 percent results from surface creep: Some of the forward momentum of the landing grains is transferred to sand grains at the surface which are simply driven ahead without being thrown into the air. A *sand storm* is intense saltation due to a high wind. In a true sandstorm the sand never gets carried more than about half a meter off the ground (because sand grains are too heavy to be taken into suspension) and the air at eye height remains quite clear. Dust storms, such as occur over agricultural lands that have become arid, result from the suspension of silt particles and represent a quite different physical process. Dust injected high in the atmosphere may travel around the world before settling out.

SAND RIPPLES

What form does a sand surface assume when wind blows over it at a speed greater than the threshold velocity? Investigation of this question will illustrate two general phenomena in physics, how an instability occurs in a physical system and how an unstable response may be damped by opposing forces until a steady state is attained.

The form of a wind-blown sand surface is hard to calculate directly, but can be

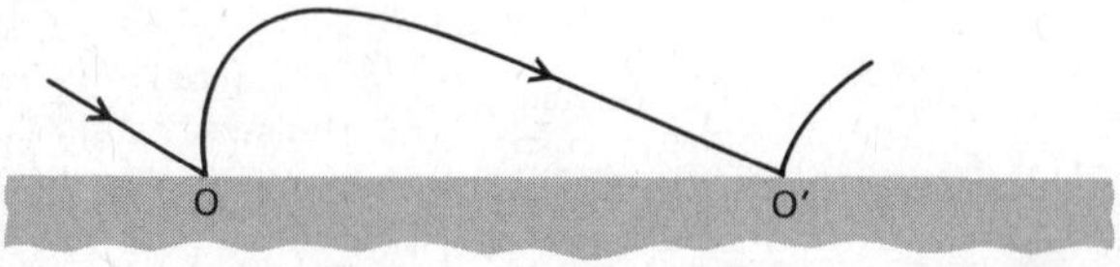

Figure 5.8 Characteristic path of sand grains in saltation. When the grain is thrown into the air at O the horizontal velocity component is zero; while in the air its horizontal velocity continually increases, causing the path to flatten. At the impact point, O′, another grain is thrown into the air, and other grains are driven ahead in creep.

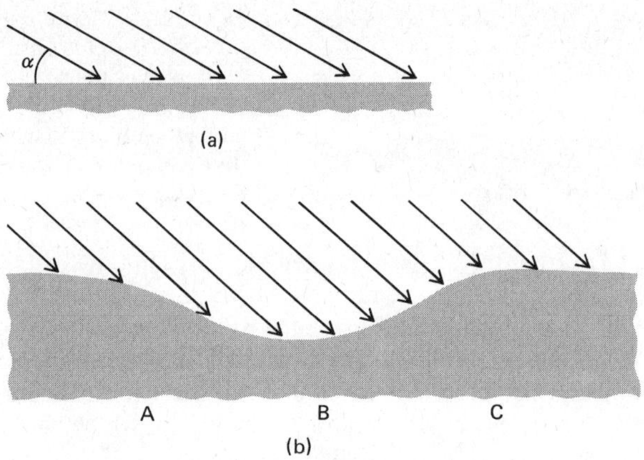

Figure 5.9 Impact trajectories of sand grains in saltation arriving on a flat surface (a) and a dimpled surface (b). The number of impacts on the leeward slope AB is much less than on the windward slope BC.

found by this procedure: Make a guess as to the surface configuration and then test whether or not this assumed configuration is stable by applying a perturbation to it. The perturbation alters the relation of the surface to the forces acting on it. If this alteration removes the perturbation, then the original form of the surface is stable. It will resist any attempt to change its form. Since no more than one surface configuration will be stable under given wind conditions, this procedure, applied to enough trial surfaces, will enable us to find that pattern.

As a starting configuration, assume a flat surface. Grains raised by saltation land on the flat surface with an inclination α (Fig. 5.9a), and are responsible both for ejecting additional grains into the air to continue the saltation process, and for driving the surface grains forward by creep. Suppose the sand surface is perturbed by a small depression as in Figure 5.9b. Is this depression stable or unstable? On the sheltered or *leeward* slope of the depression the number of grain impacts is greatly reduced while on the exposed or *windward* slope the number is increased, as shown in Figure 5.9b. Creep and the ejection of sand grains are reduced on the leeward slope, AB, but are increased on the windward slope, BC. Thus, if any small dimple appears on a flat sand surface, saltation will immediately begin to excavate more sand from the dimple; the flat surface is unstable.

This analysis suggests the true stable form of the surface. Little sand is thrown up from the leeward slope; this slope therefore casts a sand shadow downwind, starting another surface depression. Thus, a perturbation on an initially flat surface will spawn an indefinite number of further depressions downwind. A stable configuration is reached when the distance between successive crests and hollows of these depressions just matches the mean saltation distance, as shown in Figure 5.10. The stable surface form is rippled, with the ripple spacing larger, the larger the wind speed.

If no other factors came into play the ripples of Figure 5.10 would continue to grow in amplitude. But when ripples form they begin to perturb the air flow pattern

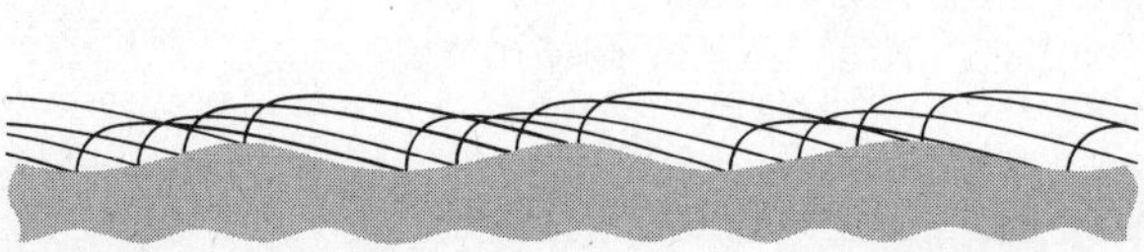

Figure 5.10 Stable configuration of ripples on a sand surface. The distance between crests matches the mean saltation distance. (After Bagnold.)

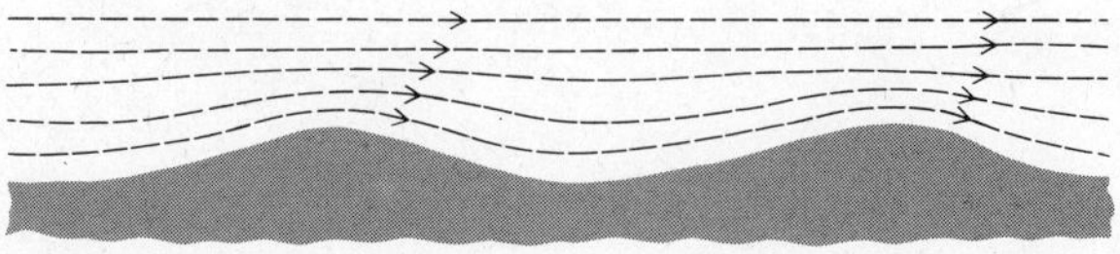

Figure 5.11 Continual growth of ripples perturbs air flow over the surface. High velocity over the crests limits the height to which they can grow. (After Bagnold.)

over the surface, as shown in Figure 5.11. Wind speed increases at the crests (where the streamlines converge); the crests then tend to be blown away. Thus, the amplitude of the ripples cannot grow indefinitely. Instead, a stable amplitude is reached when the sand shadow and wind speed effects are in balance. If the sand is perfectly uniform in size, this steady-state amplitude is quite small; if there is a distribution of sizes, the coarse, weightier, particles are left behind on the crests, which can then be built up higher, while the fine sand is found in the hollows.

These features can be studied on the dry sand above the high water mark on a beach exposed to a steady wind. Figure 5.12 shows a beach looking north with the wind blowing at force 4 (Beaufort scale, see Appendix I) from the southwest: Dry sand on the upper beach is in saltation and is being driven down over the wet sand of the lower beach in the streamers visible in the photograph. After the wind has stopped, the upper beach is covered with ripple marks (Figure 5.13). The contrast between the light and dark parts of the ripples in Figure 5.13 results from the non-

Figure 5.12 Looking north along Crescent Beach on Block Island, R. I., at low water. A Southwest wind, force 4, has placed sand on the dry upper beach into saltation; streamers of this sand are being blown down over wet lower beach.

Figure 5.13 Ripple marks on the dry upper beach of Figure 5.12 after the wind has stopped. The windward faces of the ripple marks are dark because they contain a high concentration of relatively heavy magnetite grains. The size of the ripples can be judged from the pen.

uniform composition of the beach sand. Since the photo was taken at noon, no shadows were cast by the ripples and they would not show up in the photograph if they were composed of pure white sand. Figure 5.14 is a microphotograph of a sand sample from these ripples. The dark grains are magnetite (iron oxide), which has a density nearly twice that of the light-colored grains of quartz (silicon dioxide) and feldspar (sodium aluminum silicate). The weight of the denser grains prevents them from being moved as easily as the lighter grains. On the windward slopes of the ripples the lighter quartz and feldspar grains have been removed by saltation, leaving these surfaces with a high concentration of the dark, heavy magnetite grains. The light grains concentrate on the protected, lee slopes of the ripples. As Figure 5.13 shows, the lee slopes are, as expected, the steeper ones.

DUNES

Irregularities on the surface of a sand bed can initiate the accumulation of much larger surface features called *dunes*. Small objects, such as the flat stones in Figure 5.15, trap sand in their lee. Sand carried by the wind is driven up over the object; behind it is a zone where a small eddy with lower wind speeds forms, as shown in Figure 5.16. Sand is dropped from the slow-moving air in the eddy. Larger obstacles, such as the boulder in Figure 5.17, have in front of them a zone of stagnant air which results in sand accumulation on the windward side of the obstacle. In the picture the

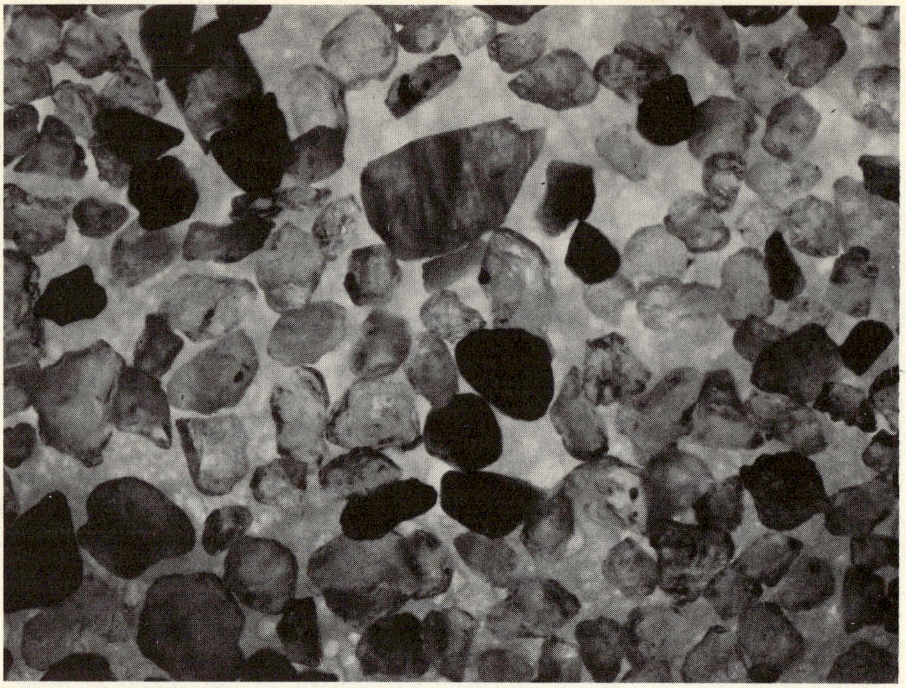

Figure 5.14 Sand from the beach of Figure 5.12 photographed at a magnification of 20 times. Dark grains are magnetite, clear grains are quartz, and the lightly colored grains showing traces of cleavage faces are feldspar.

Figure 5.15 Sand shadows forming behind small stones on the still-wet portion of the beach of Figure 5.12. Wind-driven sand is dry and so shows up as white where deposited. Size of stone in foreground is 15 cm.

Figure 5.16 Formation of an eddy under the detached boundary layer behind a small obstruction results in lower average wind speed and deposition of sand.

accumulating sand has started to spread out beyond the boulder. With continued wind such an accumulation grows into a hill which overtops the initiating obstacle and becomes a dune.

The profile of a dune is shown in Figure 5.18. Sand is driven up the windward face by the wind. Beyond the top of the dune there is a zone of turbulence and lower average wind speeds. Here the sand carried up the front face is dropped. Sand settles out at the rear face until this becomes so steep that the accumulated sand avalanches down. (The back face is called, for this reason, the *slip face*.) The characteristic shape of the dune is constant but under a steady wind, sand is continually transferred from the front to the rear face. Consequently, the dune advances in the direction of the wind. In three dimensions the dune takes on a crescent shape (Fig. 5.19) as the low edges tend to be driven ahead of the main body of the dune. Such dunes are called *barchans*.

Dunes of the ideal barchan shape are found in desert areas when the prevailing wind is from one direction. Large dunes often form in beach areas, but tend not to

Figure 5.17 At a place near that of Figure 5.15 a large boulder on the beach has blocked air flow in the forward direction and caused deposit of dry sand. Area around the base of the boulder has been scoured out by turbulence.

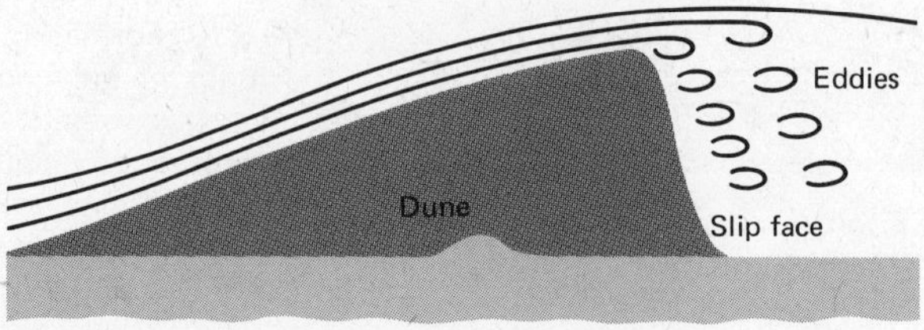

Figure 5.18 Cross section of a sand dune. The dune grows and advances by the deposit of sand on the slip face.

have the simple shape of the barchan: strong winds are likely to blow across a beach from more than one direction, and the vegetation common to beaches has a marked influence on the dune. In temperate climates beach grass grows rapidly into sand areas. Grass slows the wind speed near the surface and so impedes sand transport by saltation and creep. Because of the lower wind velocitites near the sand, beach grass is an effective sand trap; it stabilizes the dunes on which it grows. Typical beach dunes are shown in Figure 5.20; often on such dunes the slip face remains free of grass, as it receives a small supply of sand carried up from a beach. A coastal area covered by such dunes appears to be quite stable, but if the beach grass is disturbed, the dunes may be easily set in motion. Thoreau's *Cape Cod* describes the invasion of Provincetown, Massachusetts by sand released when the townspeople allowed their cattle to roam the surrounding dunes. Today dune areas near beaches are often destroyed by the wind when motor vehicles are driven through the protective cover of beach grass. The advance of dunes into a forested area near Provincetown is shown in Figure 5.21.

RIVERS

Water evaporated from the oceans is deposited on the land by rain and snow. It is removed by two processes: runoff through rivers, and evaporation (directly from the ground and through transpiration by plants). Depending on the amount and type of vegetation and the average temperature, from 10 to 50 percent of the water precipitated in a region is carried away as runoff. Rivers collect runoff water from a large area of the countryside through feeder streams, and discharge to the sea the aggregate flow. The flowing water of rivers erodes rock it passes over, and collects and removes from the land the rock waste and soil produced by erosion. A river carries the products of erosion partly in solution, partly in suspension, and partly as bed load. *Bed load* is the material too large to be suspended which is pushed along the bottom of the channel by the stream. If the stream velocity is large in relation to the load of solid materials carried by the current, the river channel will be cut more deeply by erosion of the stream bed. This increases the load carried and makes the downhill gradient of the stream less steep. If the load is too great for the current, some will be deposited until the excess load is disposed of. A river is said to be *graded* when these processes are in balance.

The interaction between flowing water and the bed over which it travels largely determines the configuration of a river channel and its course over the land. A difficulty in an analysis of river bed configuration is the large number of relevant variables. The total average discharge rate, the load of solid material carried, and the altitude difference between head and mouth are fixed by the environment in which the river is placed. Variables dependent on these conditions are channel width, depth,

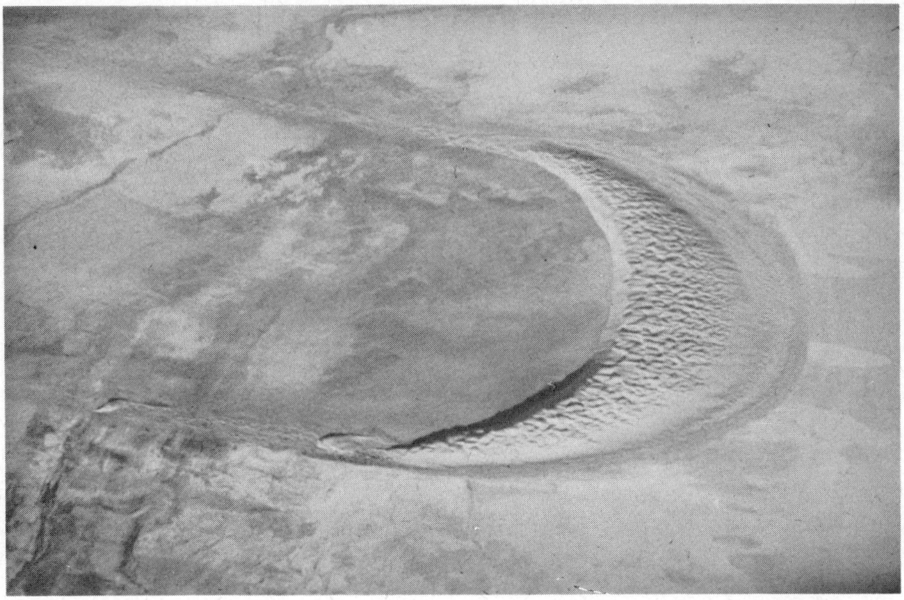

Figure 5.19 Barchans in a desert area. Courtesy of Ward's National Science Establishment.

Figure 5.20 The land at Sandy Point, Rhode Island, consists entirely of sand. The dunes have been stabilized by vegetation but a blowout has occurred in the line of dunes to the right.

Figure 5.21 Advance of coastal dunes into a forested area of the Provincelands, Cape Cod.

and slope, the particle size of the sediment carried, and the course followed by the channel over the countryside. There may be several combinations of these variables which leave the river an equally effective transportation system. The possibility of the river shifting its channel from time to time is then to be expected. Some generalizations about the form of a river channel can be made. As an illustration, consider the relation of channel plan and profile to the size of the sediment load. When the load is large and the water supply limited, a large amount of solid material can be kept moving when the stream flows in a broad, shallow channel. This maximizes the bed load that can be moved while the high turbulence in the shallow channel keeps the amount of suspended material large. But in a shallow channel the drag of the stream bed on the flowing water is large. Rivers of this type require a large slope to maintain an adequate velocity. They are found, for example, in mountainous regions (Fig. 5.22) and in the high plains of the central United States.

If the load of sediment that a river is transporting is small, a flat, shallow bed is not required to keep the sediment moving. Frictional resistance to the water flow is minimized when the channel presents a minimum of wetted perimeter; that is, when it has a relatively deep channel. Such a stream requires only a small slope to maintain its discharge rate. The approach of most large rivers to the seacoast falls into this pattern.

The course of a river is much influenced by channel form. Broad, shallow rivers tend to a braided pattern (Fig. 5.22), since during much of the year there is insufficient water to fill out the entire channel and the depth must remain as shallow as possible if the bed load is to be moved. In a deep stream the water has a much better chance to cut into the banks. The perturbation method may be used to see if a

Figure 5.22 Broad, shallow channel of the River Feshie near Kingussie in Grampian Mountains of the Highlands of Scotland. Crown Copyright Geological Survey photograph. Reproduced by permission of the Controller, H. M. Stationery Office.

straight, deep river channel is stable. Suppose that a small obstruction, a boulder perhaps, at A in Figure 5.23a deflects the stream to the left. Erosion will begin on the left bank at B while deposition will occur in the still water behind the boulder at C. A bend will begin to form in the river. But this bend will then throw the stream to the right, as in Figure 5.23b, and start localized erosion on the right bank at D. This will

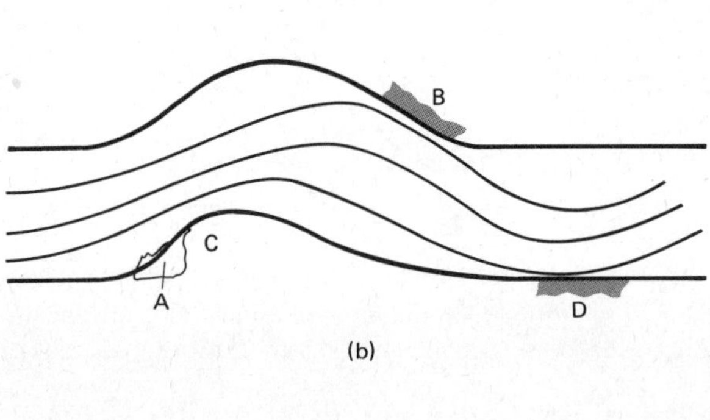

Figure 5.23 Instability of a straight river channel. An obstruction at A causes erosion at B and deposition at C. The resultant bend in the channel initiates erosion at D and so starts a new bend.

Figure 5.24 Meanders of the River Cuckmere near West Dean, Sussex, England. (Crown Copyright Geological Survey photograph. Reproduced by permission of the Controller, H. M. Stationery Office.)

start another bend in the channel further downstream. A straight channel is unstable and when the pattern of bends is fully developed the stream course *meanders*, as in Figure 5.24. Observations on many streams show that the wavelength between bends is about ten times the channel width; aside from the difference in scale, the meander patterns of large and small streams look very much the same. After a pattern of meanders is established, the meander bends migrate downstream, Figure 5.25, as erosion and deposition occur on opposite sides of each bend.

The larger rivers are important sources of hydroelectric power. In order to supply water at sufficiently high pressure to the turbines which turn the generators, a dam must be built across the river. A lake filled with relatively still water forms behind the dam. Fast running river water is slowed as it enters the lake and deposits its load of sediment in the dam basin. Over a period of years the basin is gradually filled with sediment brought in by the river. Thus the utility of any hydroelectric power installa-

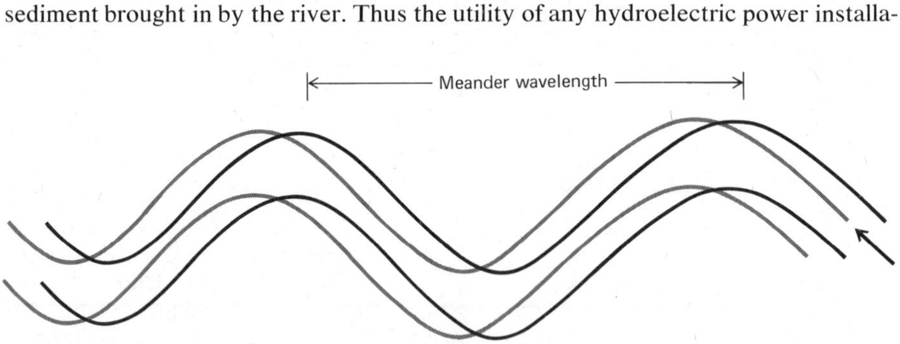

Figure 5.25 Advance of river meanders downstream as a result of a continued erosion and deposit in the pattern indicated in Figure 5.23. Initial position shown by heavy line, later position shown by gray line.

tion is limited to a finite period of time determined by the rate of sedimentation in its water basin. For Hoover Dam on the Colorado River this time is about 300 years; for the Aswan Dam on the Nile, only about 100 years. Since there is no practical way of clearing out a sediment-filled lake of large size, hydroelectricity clearly does not represent the inexhaustible resource it is commonly supposed to be.

THE PROFILE OF BEACHES

The study of wind-blown sand and of river channels deals with fluid flow in one direction over the land. Beaches, in contrast, are formed by the back and forth flow of water produced by waves. It would seem, at first glance, that the passage of waves should produce no transport of sediment over the bottom since, as waves advance over a given point, as much water moves backwards as forwards. Nevertheless, shallow water waves do transport sediment. The transport is the result of the distortion of wave form (Chapter 4) that occurs as waves move into shallow water and the nonlinear relation (Fig. 5.7) of sediment transport to fluid velocity.

As waves approach a shoaling shore, the leading edges of the crests are steepened, the crests peak up, and the troughs spread out. The water velocity over a point on the bottom then assumes the form shown in Figure 5.26. Much higher velocities flow shorewards than seawards but the duration of the shoreward flow is less (since the net water transport must be nearly zero). Suppose that the bottom is composed of uniform grains whose size makes the threshold velocity of the magnitude shown by the dashed lines in Figure 5.26. Sand is moved only during the shoreward flow, since only then is the water velocity greater than the threshold velocity. No sediment moves during the longer period of return flow because the velocity is too low. The grains are moved into the zones of breakers and then are carried onto the beach by advancing, turbulent water of the swash zone. Some grains are carried back to the breaker zone by the backwash but many are left behind because much of the swash water percolates downwards into the sand rather than returning as backwash. As a result, sand collected from the bottom offshore is carried inwards to create the typical

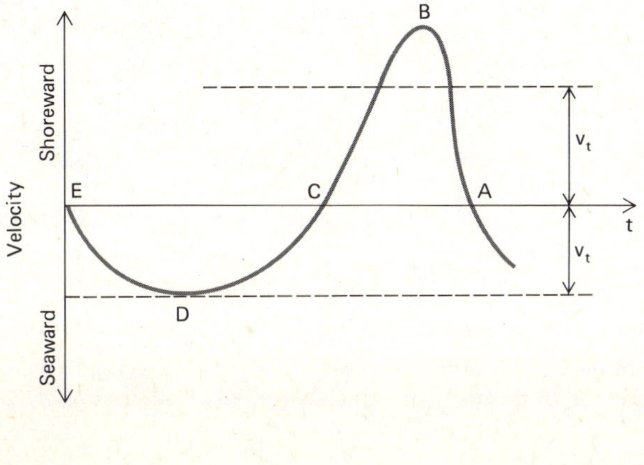

Figure 5.26 Water velocity as a function of time at a point on the bottom as a shallow water wave passes. The wave crest is steepened and shortened, producing a higher shoreward velocity. Areas ABC and CDE are equal because there is no net transport of water. The threshold velocity for the sand is v_t; it is greater for shoreward motion because the sand must be moved uphill. Sand is in motion only when $v > v_t$. The relative positions of v_t and the curve ABCDE depend on sand size and sea state.

Figure 5.27 The berm on a beach is built by the inshore transport of bottom material by waves.

beach profile shown in Figure 5.27: a smooth shoreward slope topped by a high berm which is the end point of deposition.

Such wave action leads to the formation of a beach having a limited range of grain sizes (as may be verified by examination of any sand beach, see also Figure 5.14). Larger grains have a threshold velocity greater than the highest shoreward water speed and are not moved; smaller grains are carried in suspension and move equally backwards and forward as waves pass. During a period of prolonged calm weather (typically during the summer months) moderate wave action may build up a fine sand beach. Under storm conditions the amplitude of the water velocity increases. Sand formerly deposited on the beach is taken into suspension during a storm, and commonly, deposited in deeper water (where bottom velocities are lower) as a bar, leaving behind a beach built of shingle or cobbles like that shown in Figure 5.28. At a beach where the sand supply is not abundant, a single summer storm may suffice to cause this transformation. The sand is not lost, however. It is returned to the beach during the next spell of calm weather. Beaches on exposed coasts sometimes have a

Figure 5.28 Storm Beach, Portgower, England, is made entirely of shingle (small stones). The berm can be seen clearly in the center of the photograph. Crown Copyright Geological Survey Photograph. Reporduced by permission of the Controller, H. M. Stationery Office.

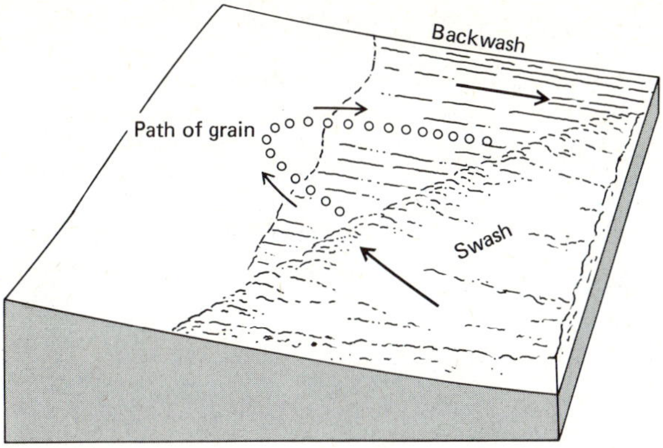

Figure 5.29 Lateral transport of sand in the swash zone due to inclined incidence of waves on a beach. From Strahler, *A Geologist's View of Cape Cod.*

completely different character in summer and winter as a result of the stronger wave action in winter.

LONGSHORE TRANSPORT

Refraction swings waves around so that they approach a beach straight on. Refraction is not always fully effective and it is not unusual for waves to reach the shore

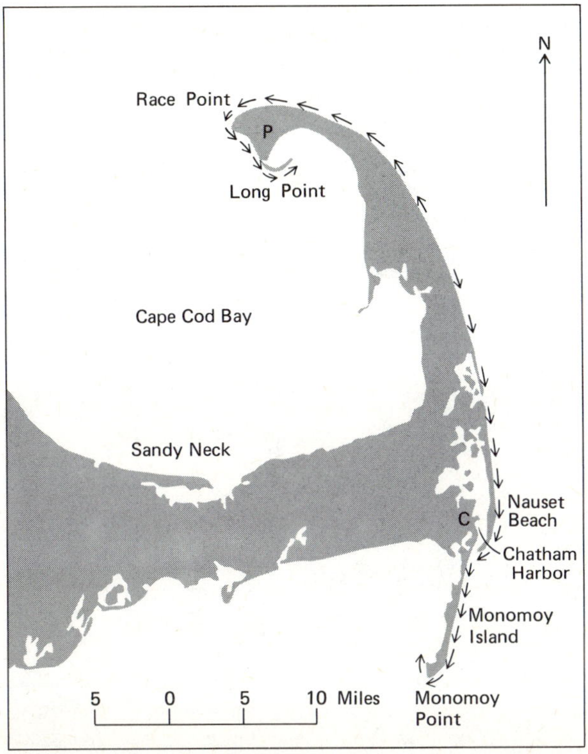

Figure 5.30 Longshore drift on Cape Cod is shown by the arrows. Areas of land built up by the drift are stippled. The Provincelands are at P. Chatham is at C. After Strahler, *A Geologist's View of Cape Cod.*

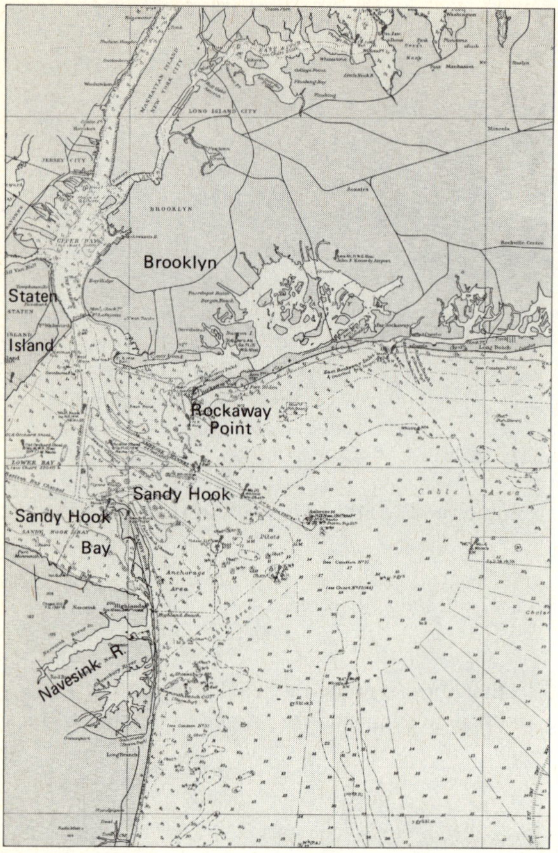

Figure 5.31 Sand spits built by longshore currents: (a) Sandy Hook and Rockaway Point at the entrance to New York Harbor; (b) on the extremities of Nantucket Island where the south and east sides face to open ocean; (c) Spurn Head at the entrance to the River Humber on the east coast of England. Reproduced from BA charts numbers 2755, 2489, 109 with the sanction of the Controller, H.M. Stationery Office and of the Hydrographer of the Navy.

(a)

(b)

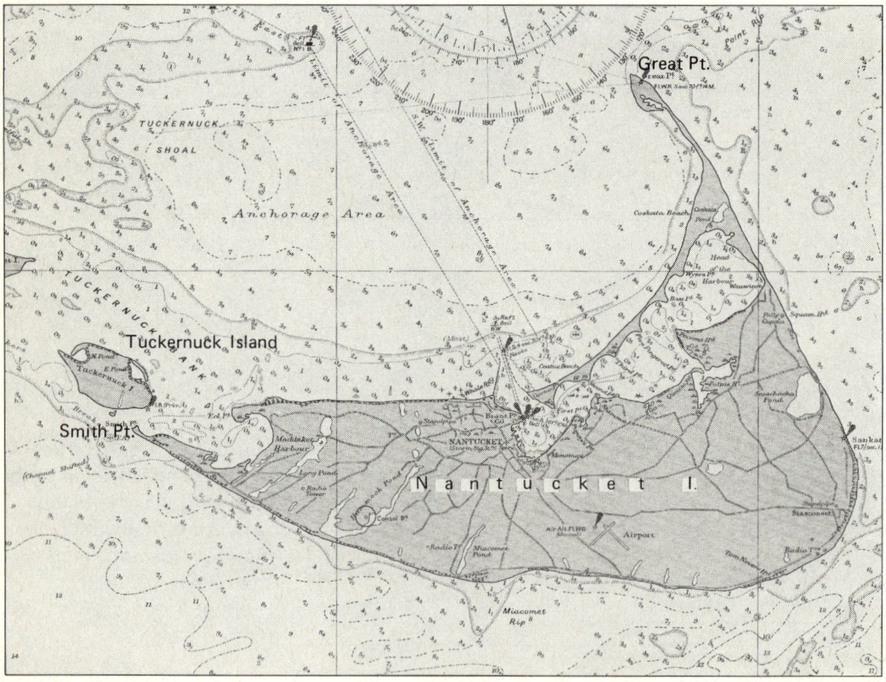

(c)

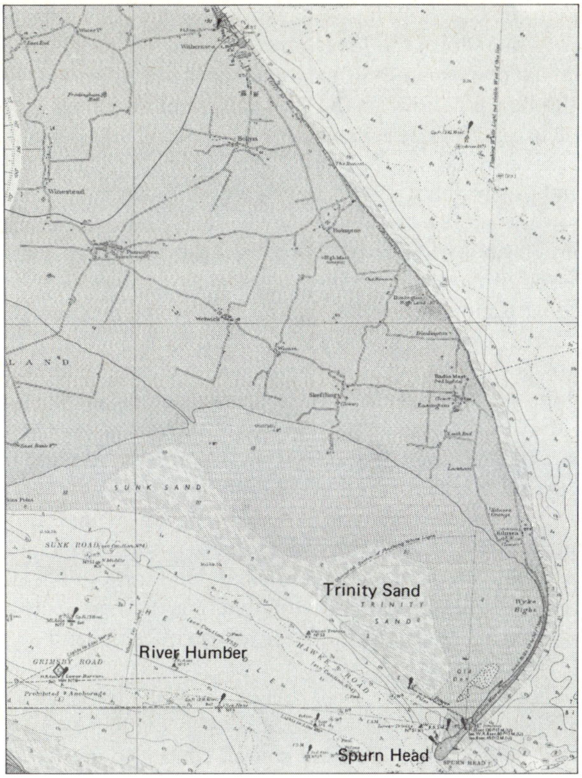

Figure 5.32 (*Below*). Erosion of the east coast of Suffolk, England, by storm waves in the North Sea. Crown Copyright Geological Survey photograph. Reproduced by permission of the Controller, H.M. Stationery Office.

with some small inclination to the beach. Water in the surf zone then develops a component of velocity parallel to the shore, the longshore current.

The most important practical consequence of the longshore current is the transport of sand along the face of the beach parallel to the shore. Lateral transport, or beach drifting, is illustrated in Figure 5.29. The longshore drift of sand may result in the building of beaches where there would otherwise be no sand supply. Alternatively, if the longshore supply of sand to a beach is cut off by the erection of a jetty or groin (a rigid structure at an angle to the shore), the beach may be lost. Shoreside communities have on occasion learned about longshore currents only after losing a beach because of harbor engineering projects undertaken some distance away.

Fine examples of the construction of beaches by the longshore transport of sand are found on Cape Cod, Massachusetts. The lower Cape, shown in Figure 5.30, consists entirely of glacial deposits rich in sand. The rise of sea level which followed the last glacial age has caused the ocean to cut into these deposits along the back (east) side of the Cape. The curvature of the Cape causes wave action to transport this sand north from the northern portion and south from the southern portion to build up the Province lands on the north and Nauset Beach and Monomoy Island on the south. Continued supply of sand to the north results in the slow growth of the sand spits Race Point and Long Point; to the south Nauset Beach is elongating at a rapid rate, making it difficult, at times, to find an open channel to Chatham Harbor. The growth of these beach areas will continue until the sand supply along the back side of the Cape is exhausted, as it must be since the cliffs there are being cut back nearly a meter each year.

Sand spits built by longshore currents are easily recognized on charts. A number of examples are shown in Figure 5.31. Their presence is clear evidence that lateral transport of beach material is occurring. The sand source can be found by looking along the coast in the direction of the prevailing winds. Owners of shorefront property in the source area are losing their land to others further along the coast who are gaining new beach. Figure 5.32 shows an area on the east coast of England where longshore currents are removing sand. The coast here has been cut back several kilometers since Roman times. The sites of numerous old villages are now well out to sea.

Another good example of how a beach is altered when the longshore supply of sand is cut off is found on Block Island, R.I. The northwest part of the shoreline of the island as mapped in 1855 is shown in Figure 5.33. The island is composed entirely of glacial deposits rich in cobbles and boulders. Consequently, beaches tend to be stony and not very satisfactory for recreation. There is, however, a large deposit of sand at the place known as Sandy Hill just north of Grace's Point on the 1855 chart. The prevailing summer winds are from the southwest. Sandy Hill is being eroded and the sand so released is being transported northwards along the west shore of the island. About ninety years ago a channel was cut from the sea into the Great Salt Pond so as to convert the latter into a harbor. Protective jetties were built along this channel; they intercept the northward flow of sand. The present configuration of the shore is shown in Figure 5.34. Sand carried northwards by the longshore current has accumulated nearly to the end of the south jetty. If it were not for the jetty, the harbor entrance would be quickly closed by drifting sand. As shown in Figures 5.35 and 5.36, a fine sand beach now stretches from Sandy Hill to the south jetty. Figure 5.37 shows the present beach north of the entrance channel. With its sand supply blocked by the jetties, it has reverted to a stony slope which is not a beach at all in the

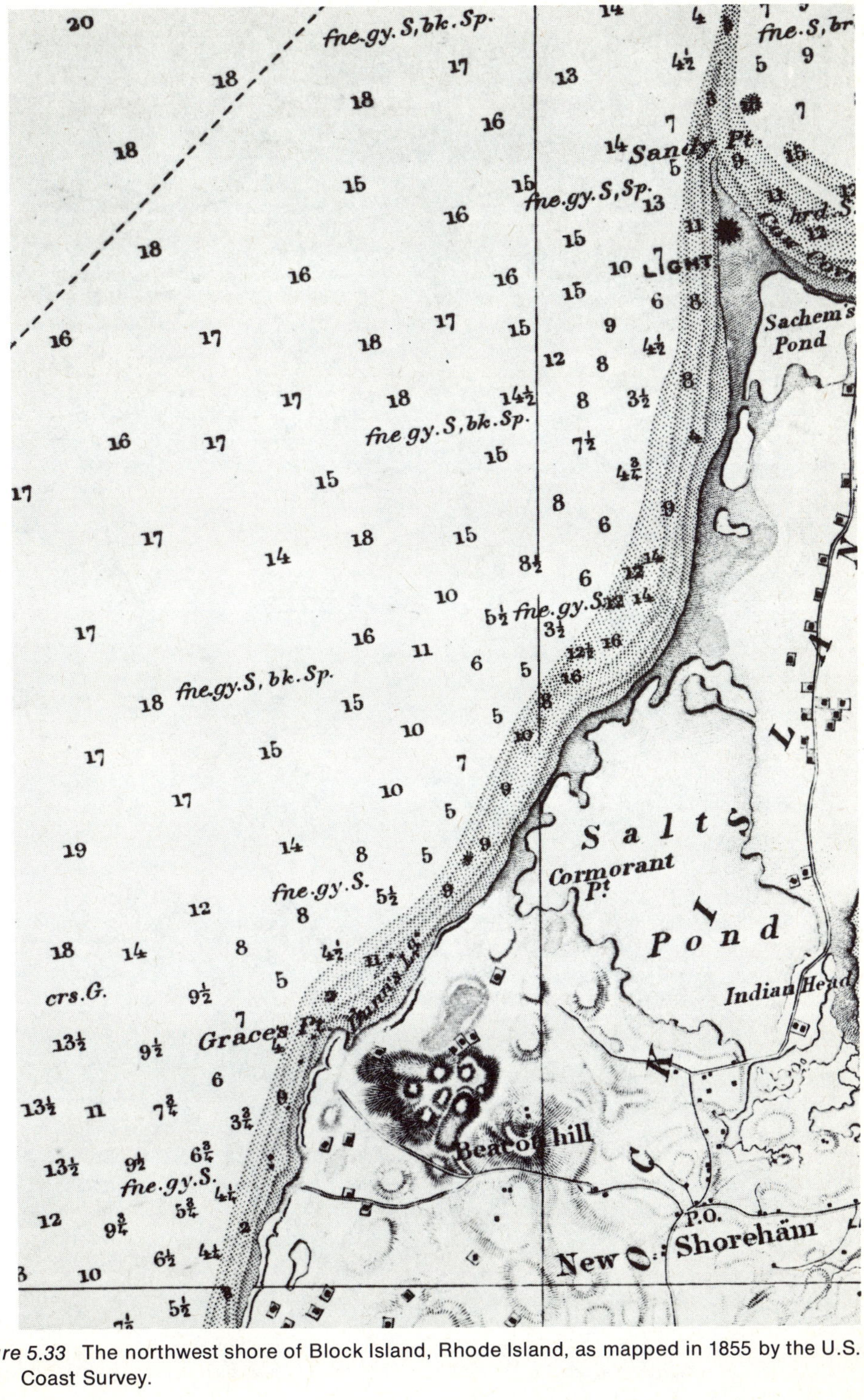

Figure 5.33 The northwest shore of Block Island, Rhode Island, as mapped in 1855 by the U.S. Coast Survey.

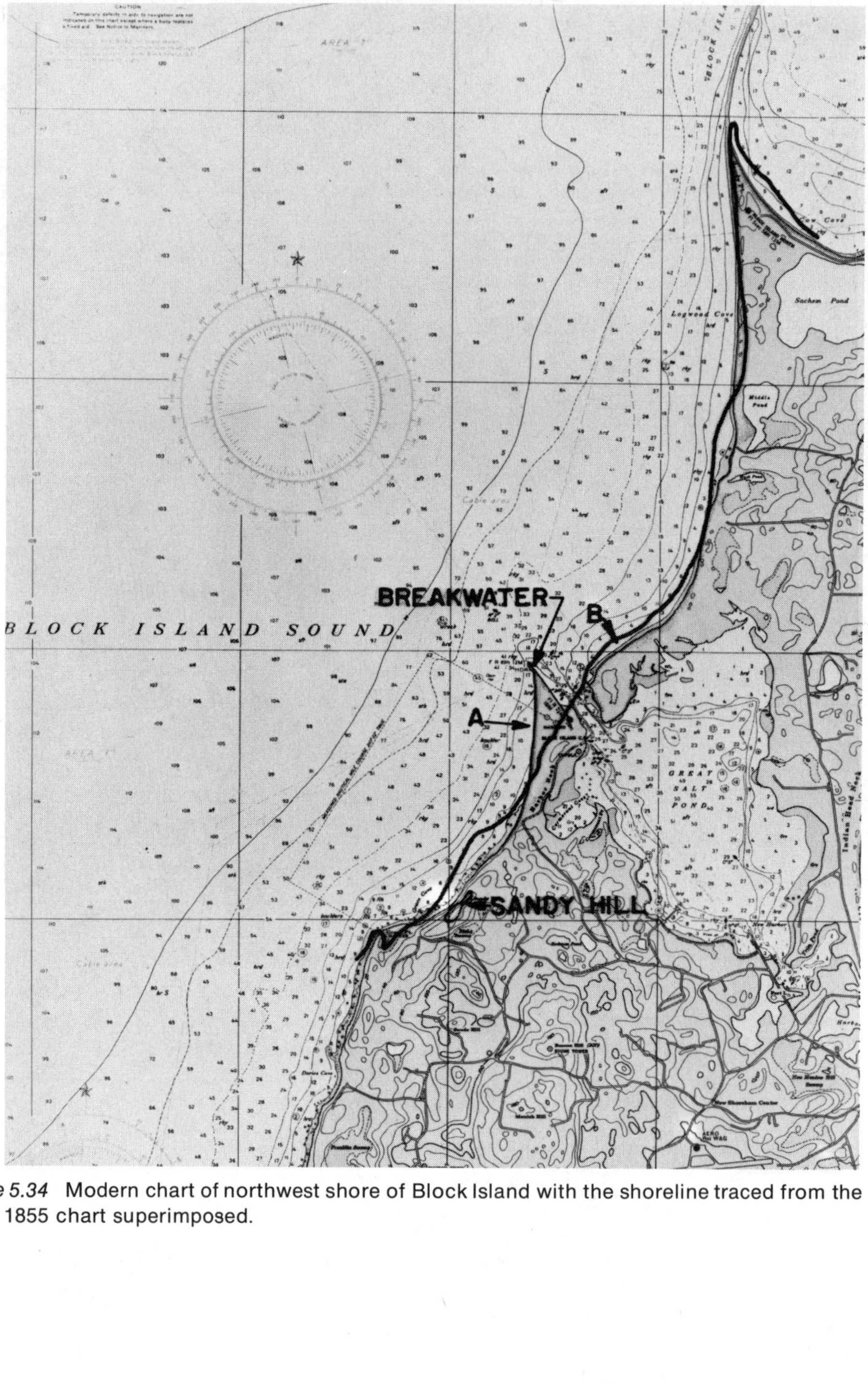

Figure 5.34 Modern chart of northwest shore of Block Island with the shoreline traced from the 1855 chart superimposed.

Figure 5.35 View at point A of Figure 5.34 looking south to Sandy Hill.

Figure 5.36 View at point A looking north. Light tower is on end of jetty. The beach is made of uniform, fine sand.

Figure 5.37 View at point B looking south towards jetty. With the sand supply cut off, only shingle remains on this beach.

recreational sense. Often, consequences of this kind are not foreseen when coastal engineering projects are undertaken. The full cost of a project in terms of damage to the property of others, to recreational facilities, and to the habitat of wildlife is then not properly assessed. Clearly, caution is in order when proposals for coastal engineering schemes are being evaluated.

Further Readings

ALLEN, J. R. L., *Physical Processes of Sedimentation*. New York: Elsevier, 1970.

BAGNOLD, R. A., *The Physics of Blown Sand and Desert Dunes*. London: Methuen, 1960.

BASCOM, W., *Waves and Beaches*. New York: Anchor Books, 1964.

BLOOM, A. L., *The Surface of the Earth*. Englewood Cliffs, N.J.: Prentice-Hall, 1969.

CARSON, M. A., *The Mechanics of Erosion*. London: Pion, 1971.

MORISAWA, M., *Streams*. New York: McGraw-Hill, 1968.

STEERS, J. A., *Coasts and Beaches*. Edinburgh: Oliver and Boyd, 1969.

ZENKOVICH, V. P., *Processes of Coastal Development*. New York: Interscience, 1967.

6 Earthquakes and the Earth's Interior

Seismology is the study of waves in the solid earth, waves whose primary manifestation is earthquakes. The unsolved problem of the control and prediction of earthquakes has an element of immediate concern about it. Apart from this practical concern, seismology is of prime interest to the solid-earth geophysicist because it is the principal source of information about the interior structure of the earth. The radiations useful as probes of outer space – light, radar, and radio signals – do not penetrate the solid earth. Only the stress waves generated by earthquakes probe deep in inner space; from the study of the propagation of these waves most of our knowledge of the earth's interior is derived.

EARTHQUAKES

As the name implies, an earthquake is a shaking of the earth. People detect the shaking in impulses transmitted through their feet (it reportedly feels like the blows of a massive hammer) and by hearing vibration of surrounding objects. These phenomena are not in themselves a hazard to the observer, but the secondary effects of the shaking may be the collapse of buildings, the rupturing of gas and water mains, and the triggering of earth slides.

Some characteristics of the earth motions in a quake can be deduced from the physical evidence left behind by a strong earthquake. Boulders on the surface may be thrown up and land some distance away without leaving marks on the intervening ground. For this to happen the quake must have produced accelerations of the ground greater than the acceleration of gravity, g. Sometimes, in a given region, objects tend to be thrown always in one direction, indicating coherent motion of the ground. Where there is abundant ground water, fountains thrown up by successive impulses indicate repeated compressions of the ground. The period of the impulses is in the range of a second to about half a minute, the pulses continuing for as long as ten minutes. Earthquake effects are strongest at a central place called the *epicenter,* and fall off gradually in intensity away from the epicenter.

These facts all indicate that an earthquake results from the arrival at the surface of the earth of waves having frequencies in the 0.03 to 1 hertz range. (One hertz = 1 cycle per second.) In intense earthquakes, accelerations of the surface are of the order of the acceleration of gravity or larger. A displacement amplitude of 10 cm is required to produce a maximum acceleration of g in a wave of approximately one hertz frequency. The waves passing through the earth need not, then, involve very large displacements. It is supposed that the waves which cause an earthquake are generated by some sudden motion—a slippage or cracking—at a place (the focus) within the earth directly under the epicenter. Away from the epicenter the wave energy is spread over greater and greater areas and so falls off in intensity.

An earthquake usually involves not one, but a series of shocks. The direction of the motions involved in each shock may be quite different. Then the residual physical evidence, being the result of a succession of events, may be confused and difficult to interpret in other than rather crude ways. To study earthquakes and wave motion in the earth, recording instruments must be used.

SEISMOGRAPHS

To detect and record the motion of the ground at the earth's surface during a quake, some reference point which remains fixed is required. It cannot be any marker rigidly attached to the ground. Ideally, it is a heavy mass suspended so as to be, as nearly as possible, free of the ground. During a quake the mass remains at rest (because of its inertia) while the ground moves relative to it. An instrument suitable for recording horizontal ground motion is the horizontal pendulum shown in Figure 6-1. The mass, M, is attached to light but rigid frame, ABC, so as to rotate freely about bearings at D and E. The axis of rotation, DE, is inclined by the small angle α to the vertical. This gives the pendulum a neutral position (the lowest point on its arc) and a small restoring force when it is displaced from neutral. The bearings D and E are joined by a

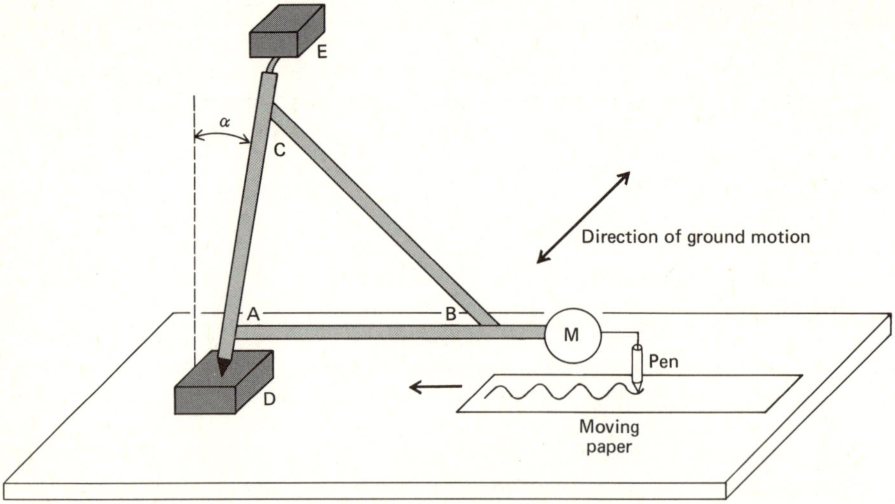

Figure 6.1 A horizontal pendulum seismograph. The mass M attached to the rigid frame ABC is free to rotate about the axis DE. The inclination, α, of the rotation axis to the vertical is small. The pen attached to M records motion of the ground on the moving paper.

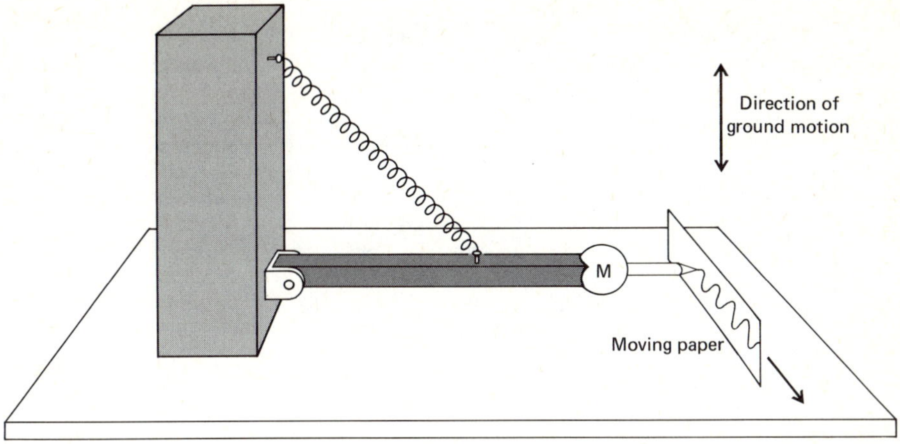

Figure 6.2 Seismograph for recording vertical motion of the ground.

rod attached rigidly to the ground. Horizontal ground motion in a direction perpendicular to the plane of the drawing displaces the support points D and E and the recording paper relative to M. The displacement is recorded on the paper by the pen attached to M; by moving the paper steadily a continuous record of this component of ground motion is obtained. Two instruments set at right angles to each other record both horizontal components of the ground motion.

To detect vertical motion of the ground, the mass and its support boom may be attached to a horizontal bearing as in Figure 6.2. A spring is used to hold the boom in a horizontal position. When the ground moves up or down, M tends to remain unmoved so that a record of vertical ground motion is recorded by the pen on the moving paper.

To be successful in recording earth motions, appropriate choices for the mechanical characteristics of these seismographs must be made. Any suspended mass has a natural period of oscillation (like a clock pendulum). The period of oscillation of the mass M must be large compared to the duration of the displacement of the earth; otherwise, M will not remain fixed during a cycle of ground motion. The seismograph must also have some damping, frictional drag on the motion of M, for without damping M would never be at rest. Obviously, the instruments must be set up in a place as free from background noise (vibrations such as those generated on highways and railways) as possible if motions due to distant earthquakes are to be recorded.

STRAIN SEISMOMETER

When the surface of the earth is displaced by a passing wave, the underlying material is strained. When a bar of material of length l_1 is elongated to length l_2, the *strain* in the bar is defined as $\delta l/l$ where $\delta l = l_2 - l_1$. A strain seismometer is an instrument set up underground to detect δl; it detects and records small changes in the distance between two reference points which are fixed in the earth. One form of the instrument consists of a long bar of fused quartz (this material has a very small coefficient of thermal expansion) fixed to the ground at point A (Fig. 6.3). Expansion or contraction of the ground between A and B causes a change in the distance between

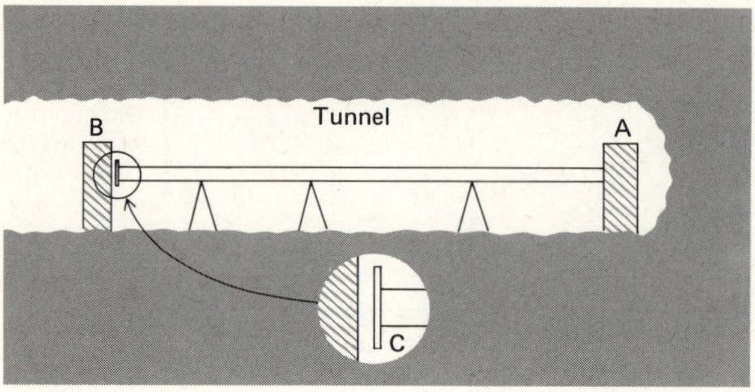

Figure 6.3 A strain seismometer. A bar of fused quartz is fixed to the ground at A. The metal plate at the other end is set close to a similar plate on the pier at B. Change in the distance between the plates is detected electrically. The instrument is set up in a tunnel to avoid displacements due to temperature changes.

the closely spaced metal plates, C. The change is detected electronically. Alternatively, changes in the distance between A and B, can be observed directly with a laser interferometer. The strain seismometer is set up in a tunnel so that it will not be influenced by strains due to heating and cooling of the earth's surface. It routinely records the strain in the solid earth caused by the tide-producing force of the sun and moon; the arrival of seismic waves is superimposed on this record.

SEISMOGRAMS

The record of wave motion obtained by a seismometer is called a *seismogram.* Seismometers may be built with any desired degree of sensitivity. The practical limit of useful sensitivity is set, not by the instruments, but by the background noise present at all observing stations. Some of this originates in human activities, but there is a continuous background of noise, much of it generated by surf pounding on the shores of the continents and some of it by unknown causes within the earth itself. Instruments of good sensitivity record signals above the background from hundreds of earthquakes each year. The sources are world-wide; the signals received from a distant source are called teleseisms. Recordings of teleseisms, such as that shown in Figure 6.4, tend to have certain common features:

The P-waves (P for *primary*) are the first to arrive and are of relatively small amplitude. Study of the components of motion of the P-waves shows them to be *compressional waves,* like the wave generated by compressing the end of a long, straight spring. In these waves the direction of particle motion is parallel to the direction of propagation. Sound waves in air are compressional waves. (The properties of waves in the earth are described more completely in the appendix to this chapter.)

The *S-waves* (S for *secondary*) are slightly larger in amplitude and follow the P-waves. They are *transverse waves:* The particle motion is perpendicular to the

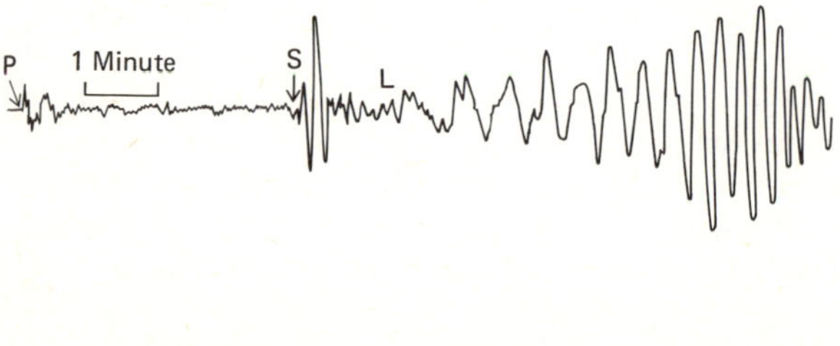

Figure 6.4 Recording of a teleseism. Arrival of the P-, S-, and L-waves is indicated.

propagation direction, as in the waves generated when one end of a clothesline is shaken.

L-waves follow the S-waves. They are of much larger amplitude than either P or S and are of substantially lower frequency (L stands for *long period wave*). The arrival of the L-waves causes the greatest destruction in a severe earthquake. The L-waves are surface waves; they travel only over the surface of the earth. This means that the energy of L-waves is confined to a much smaller volume of the earth than that of P- or S-waves. L-waves retain a large amplitude over longer distances than the P- and S-waves which are *body waves,* following paths through the earth. The P-, S-, and L-waves each have a characteristic velocity, v_P, v_S, and v_L with which they travel through the earth (see the appendix).

After the first L-waves are recorded on the seismogram, strong signals may continue to arrive which do not fall into any regular pattern for all earthquakes. They represent multiple reflections occurring within the earth. Each distinct signal on the seismogram is called a *phase*. The primary data taken from the seismogram are the arrival times of the various phases.

Besides indicating the phases and their arrival times, seismograms show the local *intensity* of the earthquake activity by the amplitude of the recorded signals. The amplitude is related to the amount of energy released by the event at the focus and to distance from the focus. An earthquake of small magnitude will produce a large intensity near the epicenter while a distant event of great magnitude produces only a small intensity. Earthquake intensity is not a precisely defined physical quantity. It depends on the number and duration of shocks, but primarily on the maximum acceleration of the ground. An intensity scale which defines twelve arbitrary levels of intensity (the modified Mercalli scale) is displayed in Table 6.1. An approximate, quantitative meaning for each intensity level is given by the values of the maximum ground acceleration. From these it is seen that the intensity intervals are logarithmic; that is, they increase by a constant factor, in this case 2.5.

The original scale of earthquake magnitude set up by Richter in 1935 is based on the greatest amplitude recorded by a standardized seismograph 100 km from the epicenter. The magnitude, M, is related to the logarithm of this amplitude, A, by

$$M = \log(A/A_0) \qquad (6.1)$$

where A_0 is an arbitrary reference amplitude chosen to set the zero point on the magnitude scale. Since it would be unusual for an observatory to be just 100 km from an epicenter, Equation 6.1 is useful only insofar as distance corrections can be applied to the observed amplitude. These are developed empirically and may be valid only in a given region. Magnitude has now been related to a more meaningful physical quantity, the total energy released at the focus, E. A useful definition is that, for surface waves.

$$\log E = 12.2 + 1.44M \qquad (6.2)$$

where E is measured in ergs. Values of M required to produce various maximum intensities are recorded in Table 6.1. At $M = 8.6$, close to greatest yet observed, $E \simeq 10^{24}$ ergs. By comparison, the total heat flow out of the earth is 10^{28} ergs/yr. This is an indication of the power available to drive mechanical processes in the solid earth; earthquakes represent a significant fraction of the work done within the earth.

Table 6.1 Scale of Earthquake Intensities with Approximately Corresponding Magnitudes[a]

Intensity	*Characteristic Effects*	*Maximum Acceleration of the Ground cm/sec²*	*Magnitude Corresponding to Highest Intensity Reached*
I	Instrumental: detected only by seismographs	1.0	3.5
II	Feeble: noticed only by sensitive people	2.5	to
III	Slight: like the vibrations due to a passing truck; felt by people at rest, especially on upper floors	5.0	4.2
IV	Moderate: felt by people while walking; rocking of loose objects, including standing vehicles	10	4.3 to
V	Rather strong: felt generally; most sleepers are awakened and bells ring	25	4.8
VI	Strong: trees sway and all suspended objects swing; damage by overturning and falling of loose objects	50	4.9–5.4
VII	Very strong: general alarm; walls crack; plaster falls	100	5.5–6.1
VIII	Destructive: car drivers seriously disturbed; masonry fissured; chimneys fall; poorly constructed buildings damaged	250	6.2 to
IX	Ruinous: some houses collapse where ground begins to crack, and pipes break open	500	6.9
X	Disastrous: ground cracks badly; many buildings destroyed and railway lines bent; landslides on steep slopes	750	7–7.3
XI	Very disastrous: few buildings remain standing; bridges destroyed; all services (railways, pipes and cables) out of action; great landslides and floods	980	7.4–8.1
XII	Catastrophic: total destruction; objects thrown into air; ground rises and falls in waves		>8.1 (maximum known, 8.9)

[a] From Holmes, *Physical Geology* (New York, Ronald Press, 1965). Reproduced by permission of the Ronald Press.

TRAVEL TIME CURVES

Suppose that in an earth made of homogeneous, elastic material an earthquake shock originates at the focus in Figure 6.5. The paths of the body waves, P and S, to observatories at A, B, and C are shown; surface waves, L, follow the circumference. If the arrival times of seismic phases P, S, and L are recorded at A, and the distance FA determined, then the velocities v_P, v_S, and v_L can be calculated. The arrival times of P, S, and L at observatories B and C may then be predicted by dividing the path length to each of these stations by the velocity. The actual arrival times for P and S at B and C are earlier than those predicted by the above calculation. This means that the v_p and v_s are greater along the paths FB and FC than along FA. The velocity must increase with depth in the earth. But if v increases with depth, the paths followed by the waves originating at F must be curved upwards by refraction. (Compare the explanation of

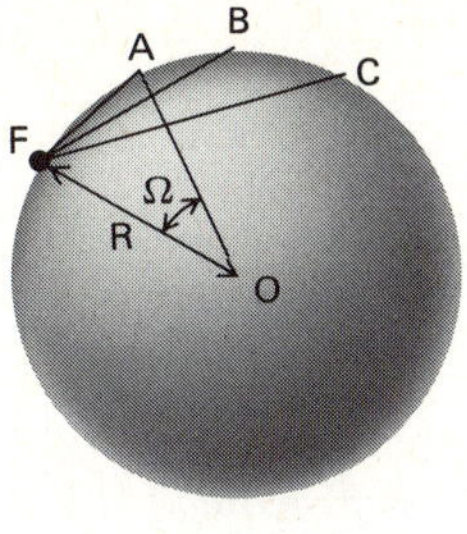

Figure 6.5 Paths of body waves from source F to observatories at A, B, and C through an earth composed of material having a constant elastic wave velocity. The distance between places on the earth's surface is conveniently represented by the geocentric angle Ω.

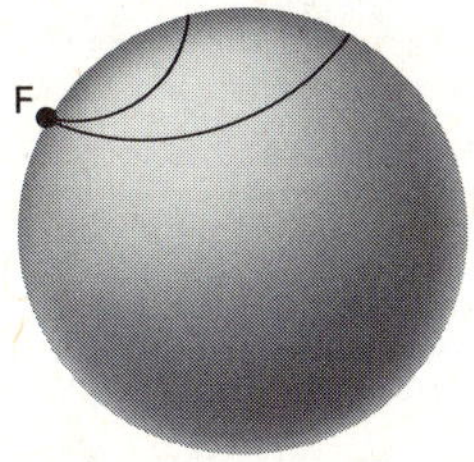

Figure 6.6 The seismic wave velocity increases with depth; body waves originating at F are bent upwards as they travel through the earth.

the refraction of waves entering shoal water given in Chapter 4 and of light in the atmosphere in the Appendix to Chapter 2.) The curved paths are shown in Figure 6.6. Since the velocity varies with depth in the earth, it is no longer possible to predict arrival times from a single set of observations, as in the example above. Instead, a large amount of observational data must be assembled so that travel time curves, as in Figure 6.7, can be constructed.

An immediate and simple use of the travel time chart is the determination of the time of origin of an earthquake and the distance of its epicenter from the observatory. Suppose that a seismogram like Figure 6.4 is recorded. The time of the originating shock and the location of the focus are not known to the observer. The procedure to find these is to plot the arrival times of P, S, and L on a strip of paper using the same scale intervals as those of the ordinate of the travel time chart. Now, slide the strip over the graph (keeping it vertical) until the three marks line up with the three curves in Figure 6.8. This will occur at only one place. The time of origin of the waves is read from the time scale on the strip at point A. The distance to the focus is read from

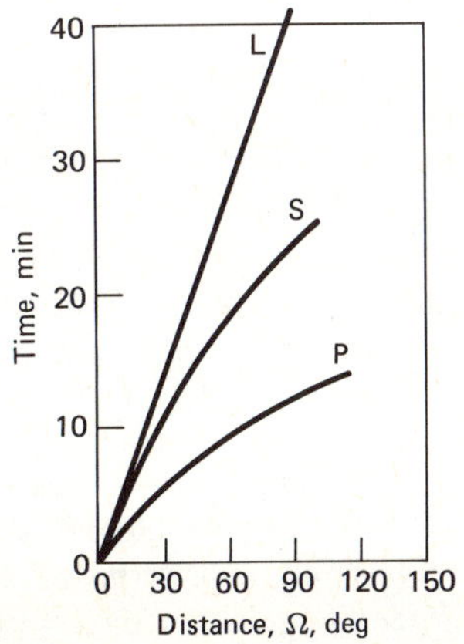

Figure 6.7 Travel time curves for P-, S-, and L-waves as determined from teleseisms recorded at many observatories. The L curve is a straight line because surface waves travel over the surface with constant velocity. The S and P curves bend downwards because v_S *and* v_P increase towards the center of the earth and the greater Ω, the deeper a ray penetrates.

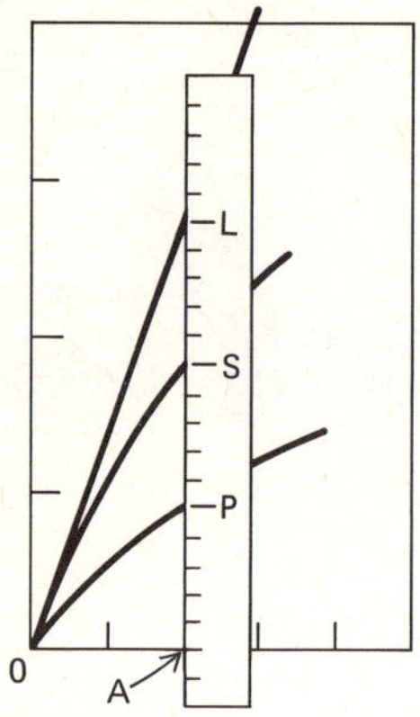

Figure 6.8 Method of finding the time of origin and the distance to the epicenter of an earthquake from data recorded at one station. The times of arrival of P, S, and L at the observatory are marked on the scale. This is then moved over the graph until a fit with the travel time curves is obtained. Time of origin is read from the time scale on the movable strip at A and the distance to the epicenter, from the abscissa interval OA.

the interval OA. If reports of the distance of the epicenter from three or more observatories are available, the position of the epicenter can be fixed at the intersection point of the distance circles centered about each observatory. By following this type of procedure seismological observatories are able to release to the press the time and place of a major earthquake even before direct reports from the afflicted region are received.

This analysis assumes that the depth of the focus below the epicenter is small. This is a good assumption for the great majority of earthquakes. A few events, known as *deep-focus* earthquakes, originate at depths as great as 700 km; for these the simple analysis above is no longer adequate.

THE CORE

There is a remarkable change in the character of seismograms recorded when the angle Ω of Figure 6.5 is greater than 105°. The P- and S-waves no longer appear on the seismogram; there is only the L-wave. Something in the interior of the earth, called the *core,* must cast an acoustic shadow on the surface. An indication of the nature of the core is given by the fact that P-waves, but not S-waves, reappear on seismograms when Ω is greater than 143°. Only the model in Figure 6.9 will fit these facts. The core in Figure 6.9 is liquid; within it v_P is low, so that P-waves are refracted downward rather than upward. S-waves are not transmitted through the core at all. A liquid core at a depth of 2900 km in the earth casts an S-wave shadow for all Ω greater than 105°, and a P-wave shadow zone from 105° to 143°. It is supposed that the core is liquid metal rich in iron: The density of an alloy of iron and nickel gives a mass for the core that is not in conflict with the known total mass and moment of inertia of the earth, and theories of the origin of the earth's magnetic field require that the core be a good conductor of electricity. Conclusive evidence as to the nature of the core is lacking and the chance of being able to sample it is probably more remote than that of men landing on Pluto.

Because seismic waves may be both reflected and refracted at the core, and reflected from the free, outer surface of the earth, the number of possible paths that waves leaving a focus may follow is large. A number of these possible paths are illustrated in Figure 6.10. Multiple reflection at the surface is indicated by repeated letters, as PP or PPP. Incidence of either P or S waves on the surface results in gener-

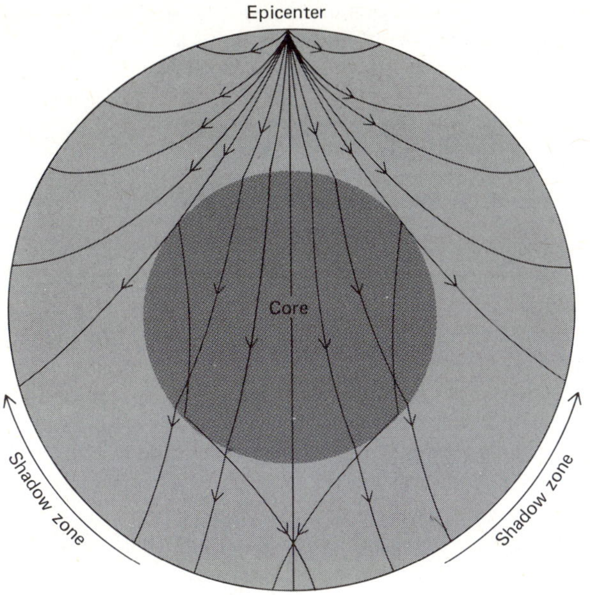

Figure 6.9 Paths of seismic waves through the earth, showing the shadow zone cast by the core. For $\Omega > 105°$ no S- or P-waves are received; P-waves only reappear when $\Omega > 143°$. This indicates that S-waves are not transmitted through the core.

ation of both a reflected P and a reflected S giving phases such as SP and PPS. Waves passing through the core are designated K, as in PKP; those reflected from the core, by a small c, as in ScS. The number of possible phases is large; the multiply reflected and refracted phases contribute to the complexity of the seismogram after the first P and S signals have been received. After the core model of Figure 6.9 was established, the phase PKIKP, which could not be accounted for in the all-liquid core

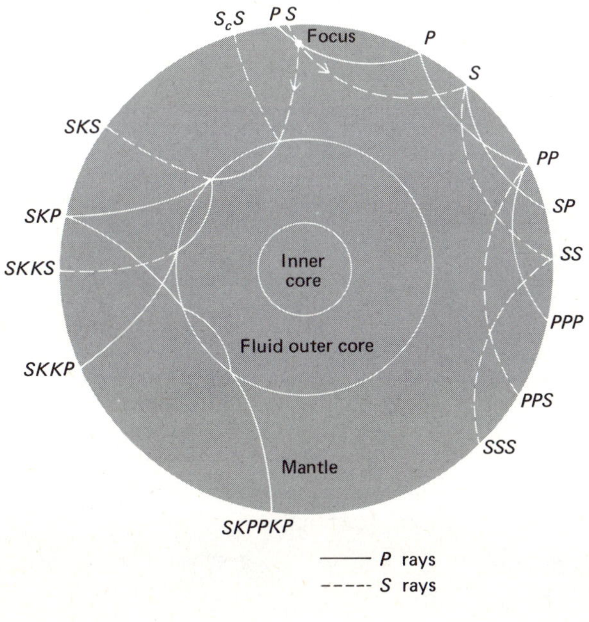

Figure 6.10 Reflection at the surface, and reflection and refraction at the core, give rise to a multiplicity of seismic signals. Some possibilities are shown here. Reflection at the surface is indicated by repeated letters; at the core, by c.

model, was discovered. This led to the suggestion that a solid inner core is present. The existence of one or more velocity discontinuities within the core is now well established, but their number and character has yet to be resolved.

THE VELOCITY PROFILE

When a good collection of travel time data has been assembled it is possible to approach the problem of finding the variation of seismic wave velocity with depth in the earth. The problem is not simple because a seismic signal follows a path that is continuously bending and along which the velocity is continually changing. Assuming that v is a function only of distance from the earth's center, R, (spherical symmetry), v for each value of R can be related to travel time and to Ω by an equation which can be solved for v only by numerical methods; this works only if the rate of change of v with R is not too great. Data from different seismic phases can be used to get cross-checks on the results. Recently determined v_P and v_S profiles are shown in Figure 6.11. When the core is reached, v_S drops to zero; there is a discontinuity in v_P because the compressional wave velocity is much less in a liquid than in a solid. The velocity rises again at the inner core boundary, suggesting a transition back to the solid state. The sharp breaks in the v_P and v_S curves in the range of depths to 1000 km are associated, it is thought, with changes in the minerals which are stable with the increasing temperature and pressure below the surface.

THE MOHOROVIČIĆ DISCONTINUITY

Analysis of teleseisms on a world-wide scale reveals the velocity profile throughout the earth and the discontinuity at the core. Study of earthquake records made at short range reveals the presence of another important velocity discontinuity, the *Mohorovičić discontinuity* or *Moho* for short, named for its discoverer, A. Mohorovičić.

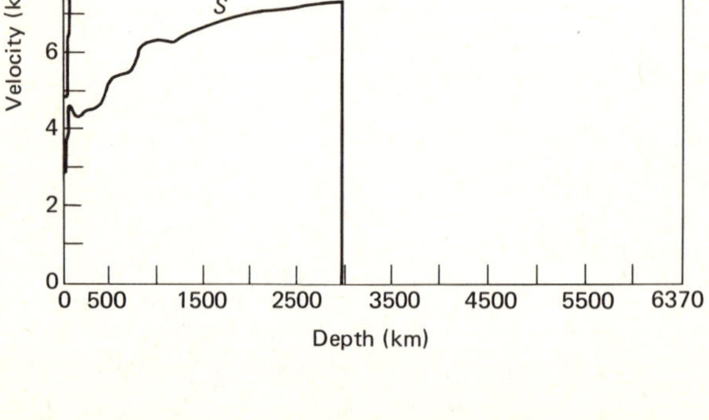

Figure 6.11 Velocity of P- and S-waves as a function of depth in the earth. The outer core boundary is at 2900 km; the inner core, at about 5000 km. S-waves are not transmitted through the outer core. After Stacey, *Physics of the Earth,* Wiley (1969).

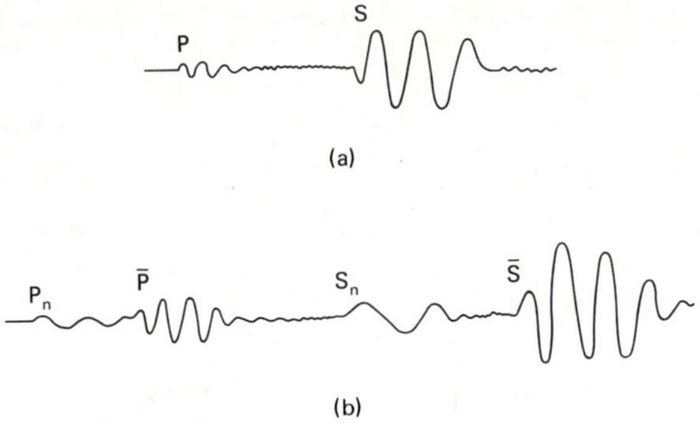

Figure 6.12 Seismograms recorded close to an epicenter (a) and some distance further away (b). The additional phases present in (b) are evidence of the Moho.

At distances of 100 to 150 km outward from the epicenter a change takes place in the character of the P and S signals received. This is illustrated in Figure 6.12. Close to the epicenter the arrival of P and S is announced by sharp, well defined oscillations on the seismogram. Further away, each arrival is preceded by a weak, relatively long period motion designated P_n and S_n. This is followed by the stronger signals $\bar{P}$ and $\bar{S}$. The travel time curves for the P-waves are shown in Figure 6.13a. The curve for $\bar{P}$ merges without discontinuity into the curve for the single P phase received close to the epicenter. The travel time curve for P_n continued out to large Ω is continuous with the P-wave recorded in teleseisms. The interpretation given to these observations by Mohorovičić is shown in Figure 6.13b. $\bar{P}$ is due to a wave traveling directly, by the shortest path, from source to receiver. P_n starts off downwards but enters a zone of higher velocity separated from the upper layer by a sharp boundary. It is refracted upwards, as shown, but travels some distance through the high velocity

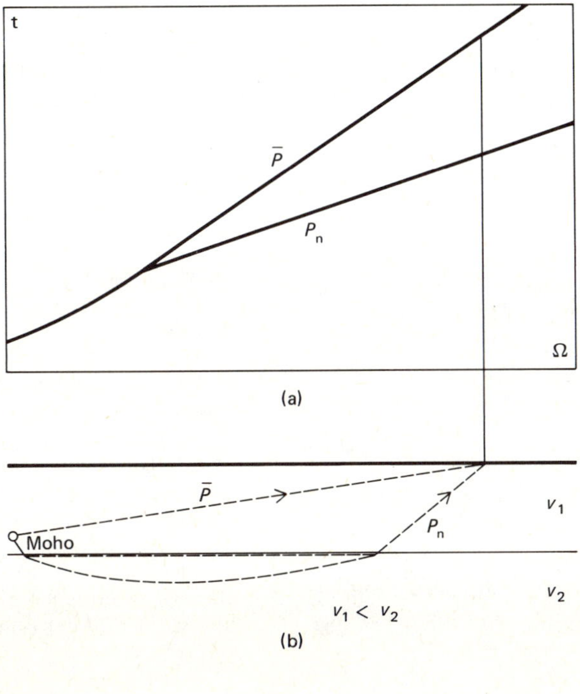

Figure 6.13 (a) Travel time curves observed for the phases identified in Figure 6.13. (b) Interpretation in terms of a two-layer structure. The upper layer of low-velocity material is separated from the material of higher velocity below by the Mohorovičić discontinuity. At large distances from the focus, waves passing through the lower layer overtake those traveling to the surface directly through the upper layer even though they follow a longer path. After Richter, *Elementary Seismology.*

material below the boundary before emerging into the lower velocity layer above. If the distance from the source is long enough, the P_n wave will reach the observer first in spite of its longer path, because a substantial part of the path is covered at higher velocity. The slopes of the two branches of the travel time curve are inversely related to the velocities in the layers; the position of the branching point is related to the layer thickness. From these data v_p for $\bar{P}$ is found to be 5.5 km/sec; for P_n, 8.2 km/sec. The thickness of the upper layer is about 30 km.

The boundary between the high and low velocity layers is the Moho. Material above the Moho is called the *crust* (a term left over from the time when it was thought that the earth consisted of a thin, solid crust floating on a liquid interior) while everything between the Moho and the core is the *mantle*. The Moho is usually interpreted as a boundary across which there is a marked difference in chemical composition. Mantle rock is thought to be rich in the dense, ferromagnesian minerals, olivine and pyroxene. (These minerals are described in the next chapter.) The crust has a much more varied mineralogy but its minerals are richer in silica and are of substantially lower density.

SEISMIC PROFILING

Seismic methods effective in studying the structure of the crust are suggested by the Mohorovičić experiment. Since distances are short it is convenient to use signals generated by exploding charges of dynamite set in shallow holes. In *refraction shooting*, detectors are set out in a line at distances R_1, R_2, and so on from the shot point (Fig. 6.14a). The arrival times of waves are plotted on a travel time curve as in Figure 6.14b. If the crust in the region being explored consists of a series of layers, each of successively greater velocity, then the travel time data can be plotted as a set of straight line segments. The velocity in each layer can be found from the slope of the corresponding segment. To determine the thickness of each layer, *reflection shooting* is used. A receiver is set up close to the shot point with the aim of detecting signals reflected back from each interface between layers of different velocity. Since the velocity in each layer is known from the refraction profile, the thickness

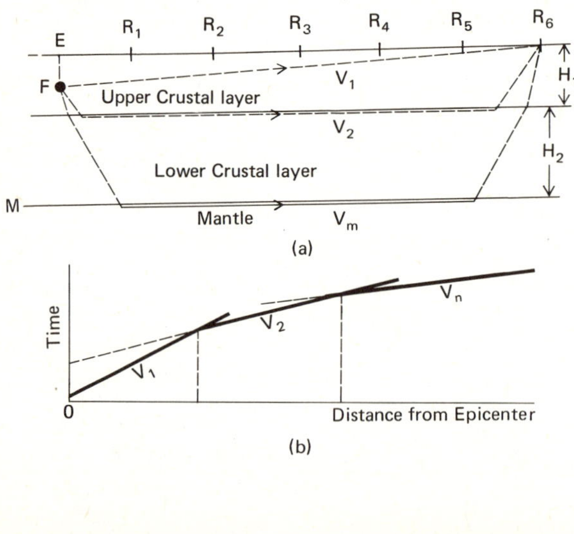

Figure 6.14 Seismic profiling on a section of crust consisting of two layers. (a) A series of receivers is set at distances R_1, R_2, and so on, on a line starting directly above the shot point. A set of rays to the receiver at distance R_6 is shown. Travel time data (b) can be fitted by a series of straight line segments. The velocity in each layer is inversely proportional to the slope of the corresponding line segment. After Holmes.

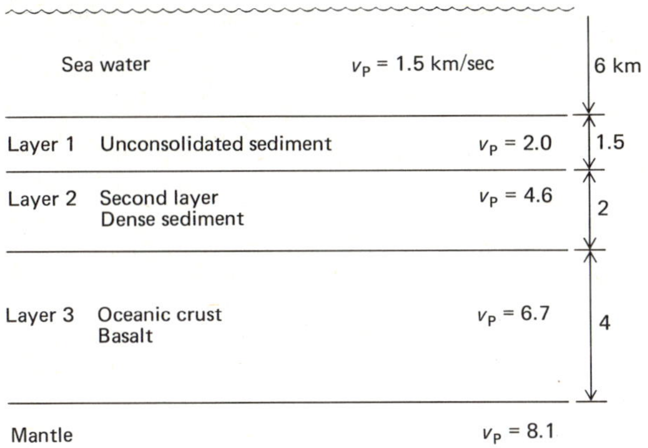

Figure 6.15 Typical ocean floor structure as revealed by seismic profiling.

of the layers can be calculated directly from the arrival times of the reflected waves.

Seismic profiling is used extensively in prospecting, particularly in the search for formations likely to be traps for petroleum. Another major use is in recording the structure of the ocean floors. At sea a line of detectors may be set out on buoys or two ships may be used. One fires repeated shots while the other, carrying receiving gear, steams away on a straight course. The first layer recorded is sea water, and, below this, various layers of sediment. Typical bottom structure revealed by seismic profiling is shown in Figure 6.15. The first layer is unconsolidated sediment, the second, densified sediment. These sediments rest on a layer of basalt, a dark, volcanic rock, about 4 km thick. Under this is the Moho. Reflection measurements show that the depth to the Moho under the ocean basins is only about 15 km. Under the continents the Moho is typically 30 km deep; under high mountain ranges it is even deeper. Thus, the crust under the oceans is much thinner than that under the continents. The explanation of this important difference will be given in Chapter 8.

EARTHQUAKE SOURCE MECHANISMS

Seimic waves may be generated by any sudden disturbance on the solid earth. They are generated by the impact of meteorites and comets, by volcanic explosions and erruptions, and by artificial explosions, particularly those of nuclear weapons. The amount of seismic energy released by all these causes is minor compared to that derived from movements occurring within the solid earth, the *tectonic earthquakes*.

The *elastic rebound hypothesis* offers one explanation of how tectonic earthquakes originate. Suppose that in some region the surface of the earth is being deformed by relative movement in opposite directions across the boundary shown as a vertical line in Figure 6.16. With the passage of time a reference line, such as AA′, consisting of points located by surveying, is distorted as in Figure 6.16b. Suppose at a later time a second reference line, BB′, is surveyed. Material near the boundary is subjected to an ever increasing distortion; this eventually becomes so large that the interface fractures, releasing the distorted rock to snap back like a compressed spring creating an earthquake. The interface is then called a *fault* because there is a discontinuity of structure across it. After fracture the displacements of the reference lines AA′ and BB′ will be as in Fig. 6.16c. Displacements of this sort are

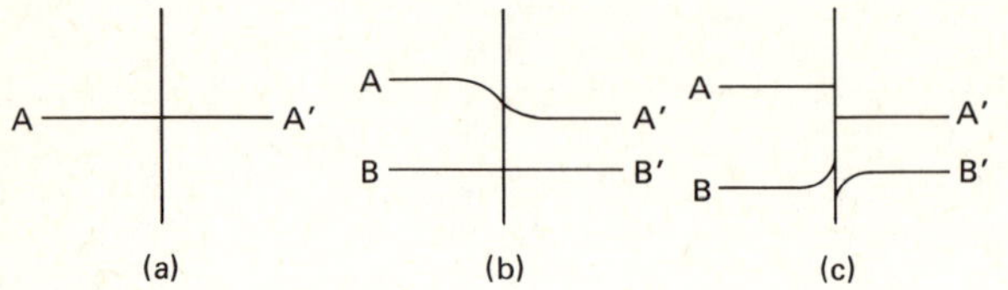

Figure 6.16 Across the vertical boundary line tectonic forces are moving the land in opposite directions. The straight line AA′ is distorted after passage of time as shown in (b) at which time another reference line, BB′, is surveyed. After release of stress by fracturing, the form of the two reference lines will be as in (c).

revealed in the alignment of surface features in Figs. 6.17 and 6.18.

In a region where tectonic earthquakes occur, a careful survey should reveal the ground motion which is responsible. In California a number of tectonic earthquakes (including the great San Francisco quake of 1906) are associated with the San Andreas fault system (Fig. 6.19). Repeated surveys across the fault system show that the land on the westward side is moving northwest at about 5 cm per year relative to the ground on the eastward side; the displacement since 1880 is a bit more than 3 meters. This movement is illustrated by the relative positions of reference marks in the Imperial Valley of California as measured in 1941 and 1954 and shown in Figure 6.20. The northwest displacement of the land in the western part of this area is clearly revealed. Because of this motion, continued earthquake activity in Southern California is inevitable.

In many earthquakes surface evidence of fault plane motion is not found. One may then question whether a fault plane exists, with the motions being obscured by overlying soil, or if some other mechanism is responsible. If seismograms from a number of observatories surrounding an earthquake epicenter are available, a test for fault plane activity can be made from examination of the direction of the first recorded motion on the seismogram. If slippage occurs on the fault plane shown in Figure 6.21, the direction of the first ground motion will be opposite in each of four quadrants surrounding the epicenter, as shown by the arrows. In fact, the recorded first motions of many earthquakes do fit this pattern. Evidently, the rebound hypothesis is applicable to many earthquakes even when no visible evidence of faulting reaches the surface. From the pattern of first motions in Figure 6.21 two possible fault plane orientations can be inferred; a decision between these possibilities must be made from other evidence, such as geological indications of the direction of tectonic motion.

In the fault plane model, distortion is released by a brittle fracture of the rock. This is a very reasonable expectation for rock in the upper part of the crust: Temperatures are low and laboratory experiments show rock to be brittle at low temperature. However, a few earthquakes (*deep-focus earthquakes*) occur at depths as great as 700 km. At that depth the temperature is near the melting temperature of the rock; locally it may be higher. Volcanoes are common in regions where deep-focus earthquakes occur. Rock at high temperature becomes quite plastic. Then tectonic displacements should be accomodated by flow of the rock without fracturing and the rebound hypothesis should not apply. Positive identification of the origin of deep-focus shocks remains to be made. Sudden, pressure-induced changes in the structure or composition of the minerals present, or some unexpected embrittlement of the

Figure 6.17 Displacement along a fault in California shown by shift in a stream valley. Courtesy Monkmeyer Press Photo Service.

Figure 6.18 Earth displacement produced by a tectonic earthquake. (U.S. Geological Survey photograph.)

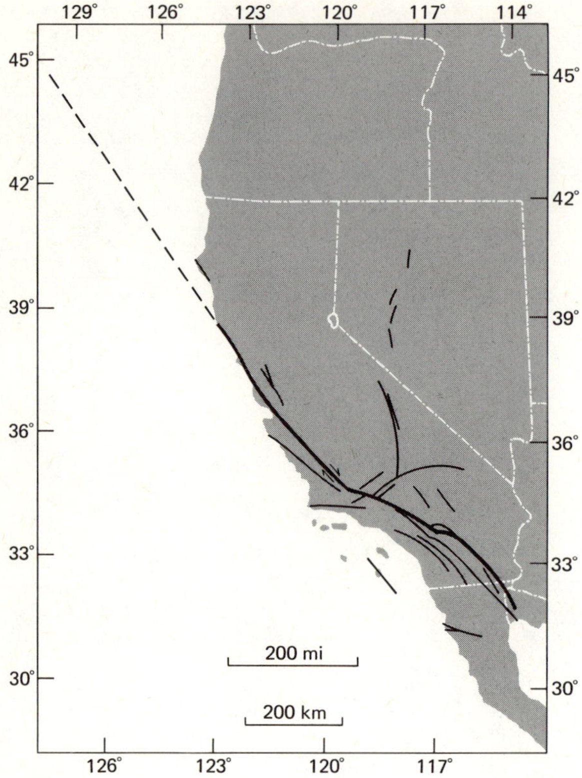

Figure 6.19 San Andreas fault system in southern California. After Richter.

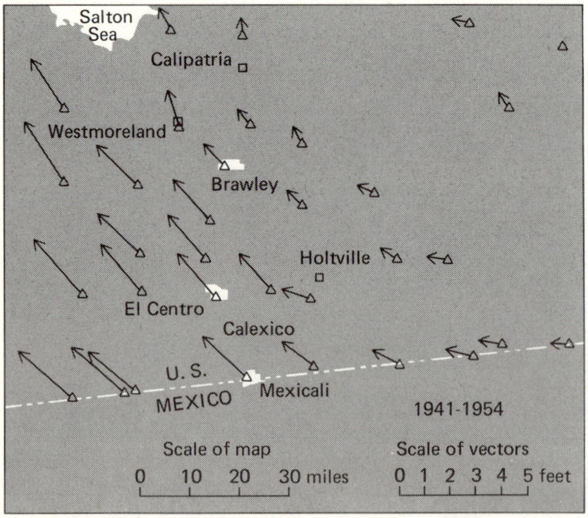

Figure 6.20 Changes in the positions of places in southern California between 1941 and 1954 as detected by repeated triangulation. The northwestward movement of the land to the west is evident. After Richter.

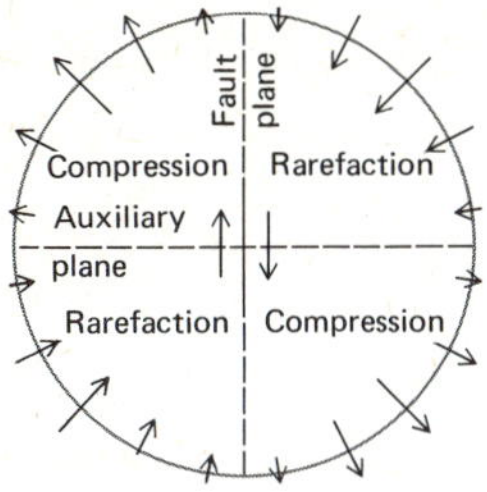

Figure 6.21 The direction of the first earth motion recorded at stations in the quadrants around the epicenter is shown for a tectonic earthquake occurring at the center of the circle. After Stacey.

rock (as by partial melting) have been considered, but definitive evidence is lacking. The identification of source mechanisms for deep earthquakes remains an important problem in seismology.

TSUNAMI

Tectonic displacements occur on the ocean floors as well as on the continents. A sudden shift in position of rock on the ocean floor generates seismic waves (detected as teleseisms on shore) and a displacement of the water above. Sudden downwards motion of a block on the sea floor causes a corresponding downwards motion of the water above. This creates an ocean wave with a wavelength comparable to the water depth, 4000 meters, say, for the open ocean. Such waves are called by their Japanese name, *tsunami.* Tsunami are shallow water waves even in the deep ocean. Their speed, given by Eq. 4.4, is $v = \sqrt{gd}$. Since d is large, this speed is very great, about 450 knots in the deep ocean. Because the wavelength is so great, passage of even a very strong tsunami would never be recognized by a ship at sea. Their arrival is, however, recorded by tide gages. For example, tsunami generated by the explosion at Krakatoa were detected on the tide gages around the Pacific and Indian Oceans. Since the time of the explosion was known, the predicted wave speed could be verified. It was as expected from Eq. 4.4.

A tsunami having no appreciable amplitude in the open ocean may be transformed into a very different wave upon approach to shallow water. The wave steepening which occurs for ordinary swell on a beach transforms tsunami into waves of enormous proportions popularly called tidal waves. (They have nothing whatever to do with the tide.) For example, on August 29, 1916, a subsidence in the floor of the Caribbean generated tsunami which reached the island of Hispaniola. Ships anchored in Santo Domingo Harbor on a clear, windless day were suddenly overwhelmed by waves large enough to dash them to the harbor bottom, about 17 meters down. (A detailed acount of this event is given in *The Wreck of the Memphis* by E. L. Beach.) Today a tsunami warning service predicts the arrival time of tsunami after the time and location of an oceanic earthquake have been determined from seismograms. This service is possible because seismic wave velocity is very much greater than tsunami speed.

EARTHQUAKE PREDICTION

A severe earthquake in a populated area inevitably results in loss of life and property. Earthquake prediction is therefore a matter to be taken seriously. While in the past

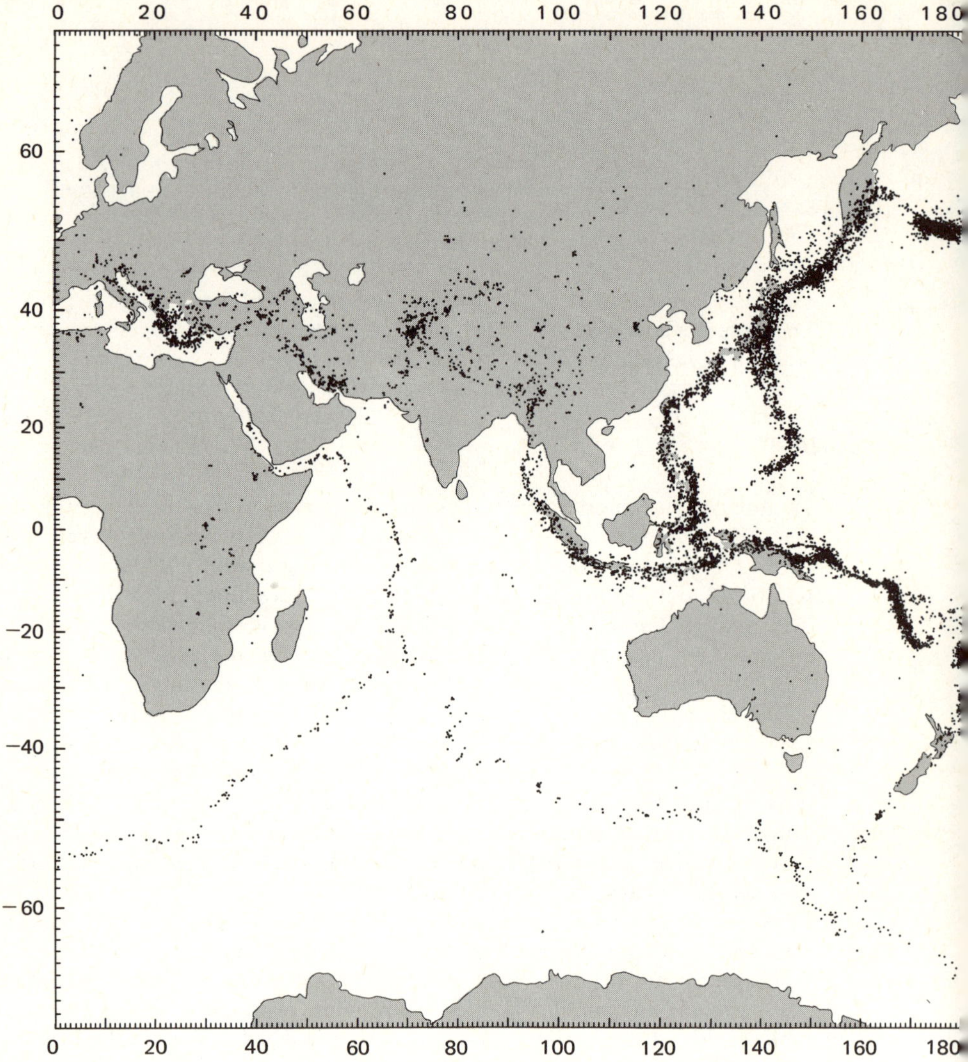

Figure 6.22 Locations of all earthquake epicenters recorded over a period of about five years. From B. Isacks, J. Oliver, and L. R. Sykes, *Journal of Geophysical Research,* **73,** 5855 (1968), based on the data of Baragangi and Dorman.

this was left to those who claimed powers as prophets, a more rational approach exists today.

One remarkable thing about earthquakes is revealed when all epicenters observed over a period of time are plotted, as in Figure 6.22: Earthquake activity is largely confined to a number of well defined bands. The belt of greatest activity surrounds the Pacific Ocean. The San Andreas fault in California is part of this belt. A major continental band of epicenters runs from southern Europe through Turkey and the Himalayas down into Indonesia. The other zones of activity all lie in the ocean basins; they follow closely the mid-ocean ridges (see Chapter 2). Outside of these ac-

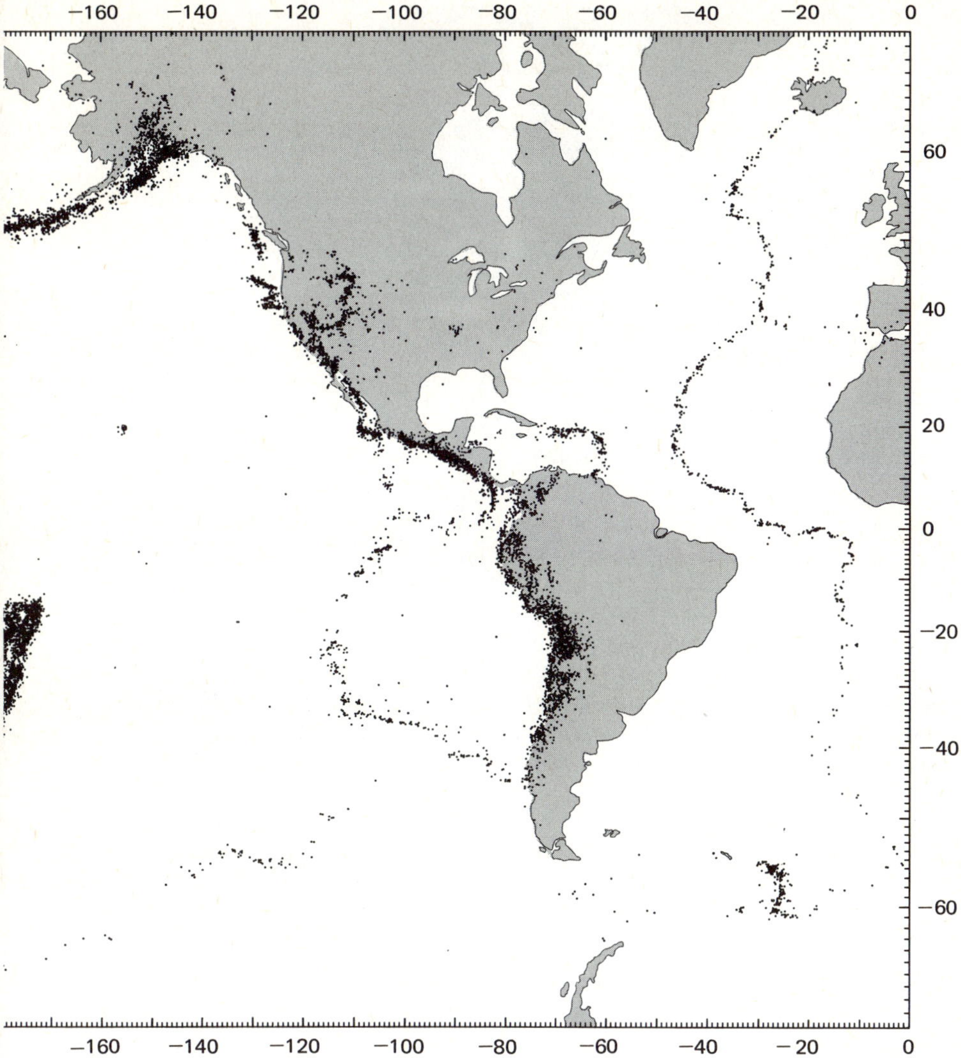

tive bands earthquakes occur, but at very much reduced frequency. Prediction will be concerned first with events within the active zones.

When fairly detailed land surveys and seismic records are available, it may be possible to determine where within a given earthquake belt large quakes are most likely to occur. In California the source of seismic activity is the northwestward motion of the land on the west side of the San Andreas fault system. The way in which this motion is accommodated along the fault differs from place to place (Fig. 6.23). Along those parts of the fault were the 1906 San Francisco and the 1857 Fort Tejon earthquakes produced breaks, the fault now appears to be locked; no slippage and no earthquake activity is detected. In the zone between Hollister and Parkfield and south of San Bernardino the fault is now active: There are frequent minor to moderate earthquakes. One interpretation is that in the active zones stress resulting from

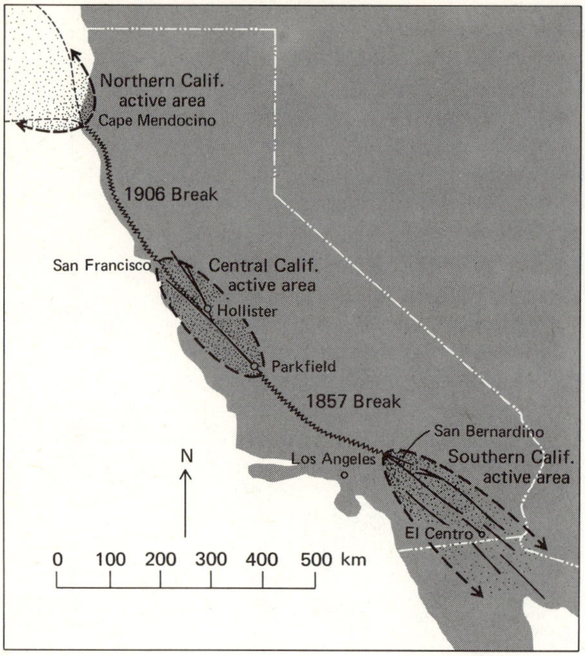

Figure 6.23 On the locked parts of the San Andreas fault system stress is presumably accumulating; in the active regions it is being slowly released by many small earthquakes. After Pakiser, Eaton, Healy, and Raleigh, *Science,* p. 1468 (1969).

the tectonic motion is being continually relieved by the small earthquakes and that along the locked sections of the fault the stress is accumulating. It can be expected to build up until released in a major earthquake. As it happens, both San Francisco and Los Angeles are near locked sections of the fault. An alternative interpretation is that stress, having been relieved by the large earthquakes, has not yet built up enough on the inactive sections of the fault to produce any seismic activity or slippage. On this hypothesis, it is in the active regions that the next big shock should occur. That a question as vital as this should remain in doubt is a clear indication of the amount of research yet to be done before even a basic understanding of seismicity in California is attained.

There is a possibility that some warning of the imminence of a major shock can be obtained if the fault is closely enough monitored. Observations in Japan have shown a marked increase in land movement (as revealed by precision surveying) a few days before major earthquake activity. It is also possible that imminence of fracture in the rock in the fault zone is indicated by changes in structure-sensitive physical properties, such as the electrical conductivity. Whether or not it would be possible to make observations of these quantities frequently enough and in enough places to have a reasonable chance of detecting an incipient break is not yet clear. Quite probably a warning of a high probability of a strong event within a period of a few weeks, or perhaps a few days, is the best that can be expected. How such a warning would be used in an urban area is also an unanswered question.

If there were some way of unlocking the locked fault segments in such a way as to gradually release the stored elastic energy, the danger of a major earthquake would be much reduced. That such a release might be possible is suggested by recent events near Denver, Colorado. Starting in 1962, waste fluids from the Rocky Mountain Arsenal were pumped down a 3.8 km deep well. Shortly thereafter a marked increase in earthquake activity near Denver was noted. Quakes up to magnitude 5.5, sufficient to produce slight damage to structures, occurred. The quakes originated in a northwest tending band at a depth of 4 to 6 km. First motion studies show that slippage occurred in a direction parallel to the active zone. Injection of fluids into the well was stopped in 1966; since then the frequency of earthquakes has dropped back to a low level. The interpretation of these observations is that a locked fault with residual stresses existed near the well site. Pumping fluid into the rock near the fault reduced its fracture strength. (This is because the pressure of the fluid supports some of the weight of the rock above so that sliding on fracture surfaces is easier.) Slippage occurred producing small earthquakes. This suggests that slow release of accumulated stress on a locked fault may be achieved by fluid injection. The technical and political difficulties which could arise in the attempt of such an experiment are easy to imagine. At present the only effective way of protecting lives and property from earthquakes is in the enactment and strict enforcement of adequate building codes in those areas where the occurrence of earthquakes is frequent. The collapse of a large, modern hospital in the 1971 California earthquake illustrates the need for more serious attention to this issue.

APPENDIX

Elastic Waves in Solids

The waves detected by seismometers are simply large-scale examples of the waves which can be generated in solid bodies during laboratory experiments. They are elastic waves because the restoring force is the elasticity of the solid (for the wind-driven waves on the ocean the restoring force is gravity). An *elastic deformation* of a solid is one which is completely recoverable upon release of the deforming force. Suppose that the bar of Figure 6.24 is compressed along its axis by force F. As F increases, the displacement of the end of the rod, δy, increases (Fig. 6.24b).

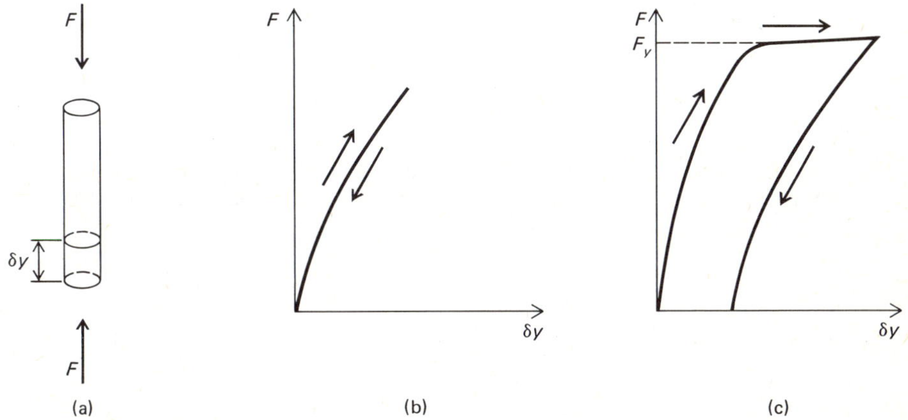

Figure 6.24 (a) Compression of the bar by the uniaxial force F produces shortening δy. (b) The force-displacement curve for elastic deformation is the same for rising as for falling stress. When F is returned to zero, $\delta y = 0$. (c) If $F > F_y$, plastic yielding occurs and, upon release of F, the bar retains a permanent (plastic) compression.

Provided the maximum compressive force is less than the yield strength of the bar, the end of the bar will return to its intial position when the force is removed. When the force exceeds the *yield strength,* permanent, plastic deformation results (Fig. 6.24c). The force amplitude in seismic waves is well below the yield strength of rock.

The resistance of the bar to elastic deformation is measured by the *Young's modulus, Y,* of the material of which it is made. This is defined as

$$Y = \frac{\sigma}{\epsilon} = \frac{(F/A)}{(\delta y/y)} \tag{6.3}$$

where A is the cross-sectional area of the bar, $\sigma = F/A$ is the *compressive stress,* and $\epsilon = \delta y/y$, the *compressive strain.* The same definition (and modulus value) holds for tensile deformation. When Y is large the material is stiff; it offers a large resistance to elastic deformation; steel has a high, and rubber a low, stiffness.

Now strike the end of the bar of Figure 6.24 a blow of short duration from above. The end of the bar is compressed against the elastic forces. The compression is propagated down the bar. When the driving impulse is released, the end of the bar expands upward. This dilatation (also opposed by elastic forces) follows the compression down the bar. A longitudinal or *compressive wave* has been generated in the bar by the blow (Fig. 6.25). The speed of the wave is high when the Young's modulus of the material is high and its density is low: When one molecule moves in such a material, a relatively large force is applied to the neighboring molecules, which displace rapidly because of their low mass. The exact relation is

$$v_t = \sqrt{\frac{Y}{\rho}} \tag{6.4}$$

where ρ is the density and v_t is called the *thin rod velocity* because it applies to compressional waves in a rod whose diameter is small compared to the wave length. In a thin rod the lateral expansion and contraction which accompany longitudinal compression and dilatation are free to occur. This is not true when the bar is thick compared to the wavelength.

A shear displacement, a displacement which distorts the shape of a block of solid, is shown in Figure 6.26. The shear stress, τ, and the shear strain, γ, are illustrated. The *shear modulus,* μ, is defined as

$$\mu = \frac{\tau}{\gamma} \tag{6.5}$$

A material which has a high shear modulus strongly resists change of shape; it has a high rigidity. A *shear wave* can be generated by applying a momentary lateral displacement to the end of a bar (or by twisting the end); its propagation speed is

$$v_S = \sqrt{\frac{\mu}{\rho}} \tag{6.6}$$

Figure 6.25 Successive compressions and rarefactions of a longitudinal, elastic wave.

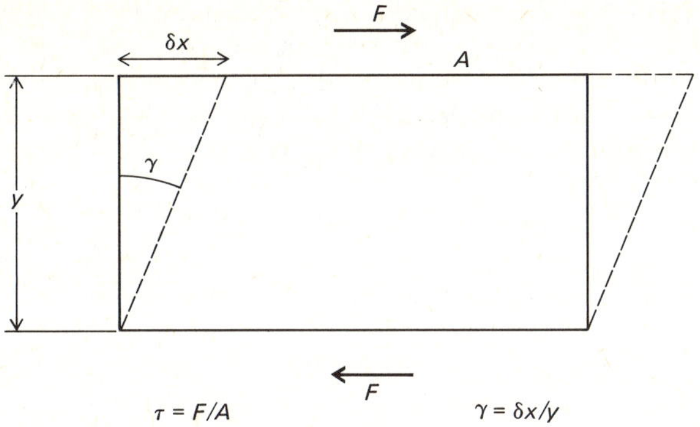

Figure 6.26 The shear strain, γ, in the block produced by the shearing forces F is $\delta x/y$. The shear stress, τ, is F divided by the surface area of the top (or bottom) face of the block.

This is also the speed of S-waves traveling through rock of shear modulus μ.

P-waves are compressional waves whose wavelength is much smaller than the size of the body they are traveling in. Because the wave does not extend to a free surface, the lateral contraction which accompanies compression and dilatation cannot develop fully. It produces a different state of deformation than that of Figure 6.24; Y is not the appropriate modulus to use for the calculation of v. The velocity of P-waves is

$$v_P = \sqrt{\left(\frac{3Y\mu - 4\mu^2}{Y - \mu}\right)\Big/\rho} \tag{6.7}$$

and in all materials $v_P > v_S$. Thus, the arrival of P before S on the seismogram is to be expected. There are no simple solutions for the velocity of surface waves in terms of the elastic constants, but elastic wave theory shows that the surface wave velocities are slightly less than v_S.

In a perfectly elastic material there is no dissipation of the vibrational energy of elastic waves; a bell made of perfectly elastic metal would, when isolated, ring forever upon being struck. The elasticity of all real materials is imperfect; vibrational energy is dissipated by *internal friction* acting within the material itself. Once excited, oscillations in a vibrating body decay with time as in Figure 6.27. The magnitude of the internal friction is measured by the decrement, Δ, defined as

$$\Delta = \log \frac{A_n}{A_{n+1}} \tag{6.8}$$

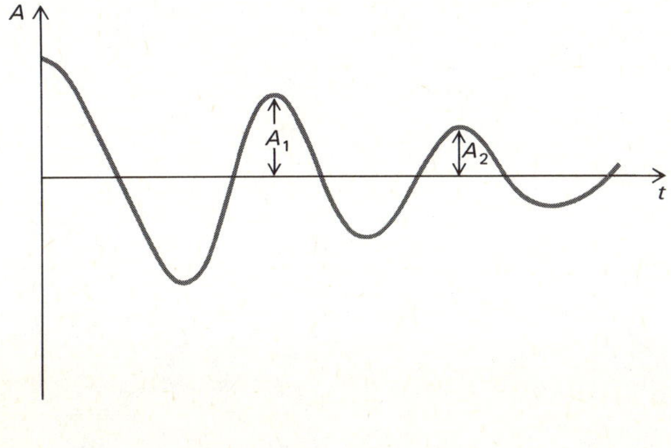

Figure 6.27 The vibration of an object made of material having imperfect elasticity decays because of internal friction. The amplitudes of successive vibrations are A_1, A_2. . . .

where A_n and A_{n+1} are the amplitudes of successive oscillations. Laboratory experiments show that Δ for rock is usually of the order of magnitude $10^{-2} > \Delta > 10^{-4}$, small enough that seismic waves can be propagated over long distances, with the eventual absorption of their vibrational energy after a few reflections within the earth.

Further Readings

BEACH, E. L., *The Wreck of the Memphis.* New York: Holt, Rinehart and Winston, 1966.

BULLEN, K. E., *An Introduction to the Theory of Seismology.* 3rd ed., New York: Cambridge, 1965.

Earthquake Prediction: a Proposal for a Ten-Year Program of Research. Office of Science and Technology, Washington, D.C., 1965.

RICHTER, C. F., *Elementary Seismology.* San Francisco: Freeman, 1958.

STACEY, F. D., *Physics of the Earth.* New York: Wiley, 1969. [Chapter 4]

7 Inside the Solid Earth

The occurrence of earthquakes in sharply defined belts around the world shows that the solid earth is not quiescent, that some pattern of motion is present in the rock below the surface. Geologic structures exposed by erosion often reveal folding, bending or cracking of the crust. Figure 7.1, for example, shows sharp folds in originally flat layers of rock. It was shown in Chapter 5 that had not new mountains been pushed upwards, erosion would long ago have reduced all the continents to featureless planes. What energy source drives the motions of the solid earth? How can solid, rigid rock be grossly deformed? It is with these questions that this chapter deals.

THE EARTH'S HEAT

Within the earth there is only one source of energy large enough to accomplish the prodigious work of forming a mountain range—the earth's heat. Although suggested by volcanic action, the earth's heat only begins to be felt some distance below the surface. Going down a mine shaft, the temperature continually rises; in fact, the principal factor limiting the depth to which mines can be pushed (as for gold in South Africa) is the high temperature encountered deep underground. The rate of increase of temperature with depth is called the *geothermal gradient.* It amounts to 10 to 30° C/km except close to regions of volcanic activity where it is locally higher.

The increase of temperature with depth in the earth shows that there is an outward flow of heat from the earth's hot interior to its relatively cool surface. The amount of this heat flow depends on the thermal conductivity of the rock and the temperature gradient. The *thermal conductivity* is a measure of how easily heat flows through a material. In the earth it depends principally on the minerals present and the rock structure. Some simple observations indicate that the thermal conductivity of the earth is low: Daily variations of surface temperature are hardly detected at a depth of one meter and the changes which do penetrate arrive half a day or more late. At a depth of 1 km evidence of low temperature remaining from the last glacial age (10,000 years ago) is found.

Figure 7.1 These originally flat-lying layers of sedimentary rock near Saundersfoot, Pembrokeshire, have been sharply folded as a result of compressions in the earth's crust, then exposed to view by erosion. Crown Copyright Geological Survey photograph. Reproduced by permission of the Controller, H. M. Stationery Office.

Because the thermal conductivity is small, the outward flow of heat from the earth's interior is also small. Indeed, it is too small to detect by direct measurement; it must be calculated from measurements of the earth's thermal conductivity and temperature gradient. An instrument for making heat flow measurements where there is soft, undisturbed sediment, as on the sea floor, is illustrated schematically in Figure 7.2. After the probe is inserted, the temperature gradient in the sediment is found from the temperature measurements along the probe. Then electric heating is applied to the bottom of the probe at a known rate, and after a steady state is reached the temperature gradient is again measured. The thermal conductivity of the sediment is calculated from the second set of observations; then the heat flow is found from the

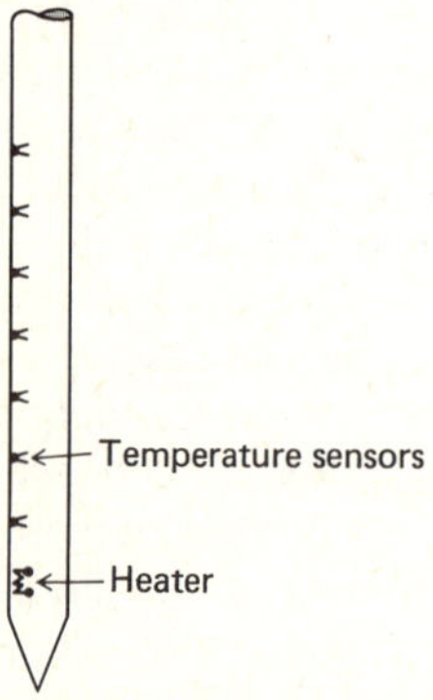

Figure 7.2 Probe for determining heat flow in soft sediments of the sea floor. Spaced along the probe are a series of electrical temperature sensors; at the bottom is an electric heating element.

first-measured temperature gradient. On land, where there is not a homogeneous, undisturbed sediment layer in which to measure, mines and bore holes are used to find the temperature gradient, and direct laboratory determinations are made of the thermal conductivity of rock samples.

Measurement confirms that the heat flow is indeed small, averaging 1.4×10^{-6} cal/(cm^2 sec). In more familiar terms, this is only enough heat to melt a layer of ice 5 mm thick in a year. Thus, the geothermal heat is small compared to the energy arriving from the sun, too small to have any direct effect on climate. Observations also show a characteristic variation in heat flow between different geologic provinces, as shown in Table 7.1. The residual heat of volcanic activity is evident; the other variations must be explained by any model of earth dynamics.

Table 7.1 Observed Heat Flow in Various Geologic Provinces

Area	*Heat Flow*
Stable continental areas ("shields", no deformation of crust detected)	0.9×10^{-6} cal/(cm^2 sec)
Areas having active volcanism in the last 70 million years	2.2
Average on continents	1.4
Ocean ridges	1.8
Ocean trenches	1.0
Average of ocean basins	1.4

If the average total heat flow from the earth is divided by the earth's mass, the heat reaching the surface per gram of earth weight per year is found to be 1.7 ergs/(gm yr). This much power would be sufficient to lift a gram of material at the surface at the rate of 0.002 cm/yr, and more than enough to raise mountains faster than the average continental erosion rate can wear them down (remember that mountain formation is local rather than continuous). The heat energy of the earth is sufficient to drive all *tectonic processes,* the processes involving deformation of the crust.

What is the source of the earth's heat? Some of it may be left over from the formation of the earth; the gravitational work done in assembling and compacting a body from small pieces is large and would appear as heat. Another source of the heat is the decay of the radioactive elements which occur naturally in rock. Radiation emitted by these elements is absorbed in the surrounding rock and converted to heat. The most important radioactive elements in the earth are U^{235}, U^{238}, Th^{232}, and K^{40}. During solidification of rock, fractional crystallization tends to concentrate these elements into silica-rich crustal rock, such as granite (see *The Chemistry of the Earth* by K. K. Turekian). The radioactive heat production in different rock types is given in Table 7.2. The data in the table suggest that (depending on the abundance of granite and basalt) a substantial part of the geothermal flux could be of radioactive origin. They also pose a problem: Granite occurs on continents and not in ocean basins, yet Table 7.1 shows that the continental and oceanic average heat flow is about the same.

Table 7.2 Radioactive Heat Production

Crustal Rocks:	Granite	360 ergs/(gm yr)
	Basalt	53
Peridotite (typical mantle rock)		0.3

GEOTHERMAL POWER

The figures given above show that the average flow of heat out of the earth is too small to be a significant source of commercial power. But in some places in the crust, usually near regions of recent volcanic activity, large reservoirs of ground water have become heated to temperatures above 100°C. Hot springs and geysers are often found in these areas. If such a reservoir of hot water can be tapped, the heated water on being brought to the surface turns to steam (because the reduction in confining pressure lowers the boiling point to 100°C). This steam can be used to turn a turbine and so generate electricity. Geothermal power installations are in operation in a number of places around the world, such as Larderello, Italy (370 megawatts) and Wairakei, New Zealand (290 megawatts). Like tidal power, geothermal power is a pollution-free source of electricity, but unlike tidal power, it is not an inexhaustible source. Once the existing reservoirs of hot water are used up, the geothermal source is exhausted. The regeneration time from heat flow in the earth is too long to make continuous steam production possible. If all known geothermal sites in the world were utilized to generate electricity with the steam being withdrawn over 50 years, the average annual geothermal power production would be about 60,000 megawatts. This is comparable to the total potential tidal power production. In terms of the total power requirements of the industrialized nations it is an insignificant contribution: Geothermal power will not be the solution to the problem of providing a large scale source of pollution-free power.

TEMPERATURE AND PRESSURE IN THE EARTH

To study the dynamic processes occurring in the solid earth it is important to know the earth's temperature gradient at least to the base of the mantle. Unfortunately, it is not possible to make the observations needed to fix the temperature distribution within the earth and the temperature gradient remains one of the most elusive geophysical quantities. Two known facts put bounds on possible gradients: First, the mantle is nearly all solid, since it transmits S-waves. Throughout the mantle, the temperature must be below the melting point of mantle rock. Second, the outer core is liquid, the inner core solid. The interface between them must be at the melting point of the core alloy.

These are not powerful constraints because there is no certain knowledge of the composition of either the mantle or the core. Even for rock of known composition, the melting point varies with depth because of increasing pressure. Most solids expand upon melting; to melt, the material must do work against the external pressure. Hence, as pressure increases, so does the melting temperature.

It is supposed that the core is composed predominantly of iron and that the inner core/outer core boundary is not the result of a composition change. The pressure at a depth of 5000 km in the earth raises the melting temperature of iron to about 3725°C (it is 1535°C at the surface). This, then, fixes one point on the temperature profile. The only other fixed data are the surface temperature and the initial slope of the temperature-depth curve. These bits of data are shown in Figure 7.3. What sort of curve passes between these points and under the mantle melting line depends on the variation with depth of the concentration of radioactive elements, of the thermal conductivity, and of the amount of heat transfer by radiation. It will also depend on the occurrence of mechanical motions (for example, convection, which accelerates heat

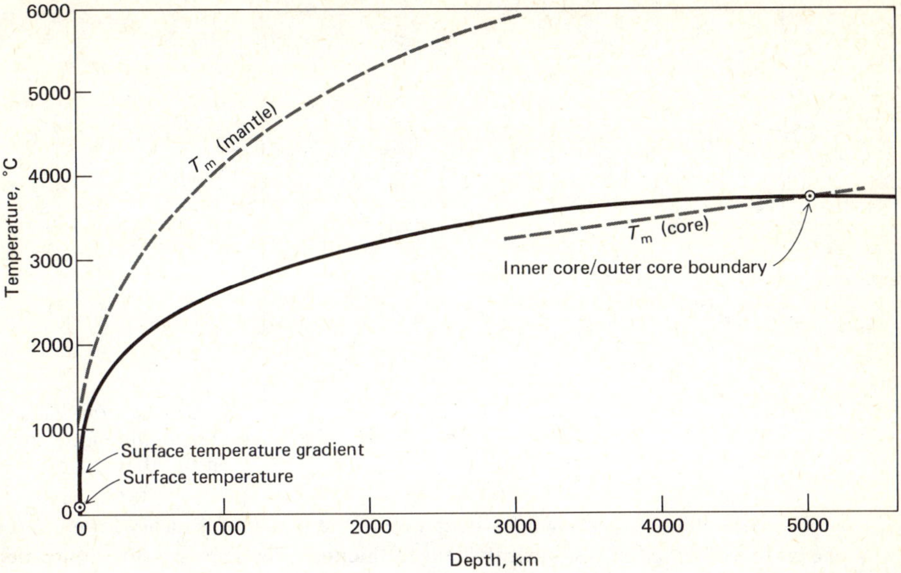

Figure 7.3 Data which the calculated temperature profile of the earth must fit are the surface temperature and gradient, and the temperature at the inner core/outer core boundary. The estimated melting temperature, T_m, for mantle and core is also shown. In the mantle $T < T_m$; in the outer core, $T > T_m$. The solid line representing the best estimated temperature profile of the earth is fitted to these constraints.

transfer in the mantle and core). Since there is no reliable knowledge of any of these quantities, the temperature profile cannot be fixed. The dashed curve in Figure 7.3 is based on current best estimates of the unknown factors. Its limitations should be kept in mind.

Determination of an approximate pressure gradient in the earth presents less of a problem. First it is necessary to find how density varies with depth. An approach to this problem can be made by utilizing the fact that the rate of change of density with depth at any point in the earth depends on two factors, the elastic wave velocities v_S and v_P, and the density at that depth. The velocities v_S and v_P are known for all depths, but density must be found. This can be done if the density is accurately known at some level near the earth's surface. At this level the rate of change of density can be found from the above relation and used to estimate the density at a slightly greater depth. The procedure is repeated over and over until the density is determined at all depths. If the earth were chemically homogeneous, the density distribution so obtained would be exactly correct. However, there is reason to believe that composition changes will be met on going down into the earth. The density distribution must be adjusted by trial to allow for these. Two constraints limit the possible choices: the mass of the earth computed from the density and volume at each depth must equal the observed mass, and the distribution of mass must give the earth's moment of inertia, its resistance to angular acceleration, which can be determined independently from astronomical observations. Within these constraints there is little latitude for choice in the density profile; the accepted curve is shown in Figure 7.4. The density contrast between mantle and core is evident. The densities are in accord with a model in which the core is an iron-nickel alloy and the mantle is composed of silicate rock rich in magnesium and iron.

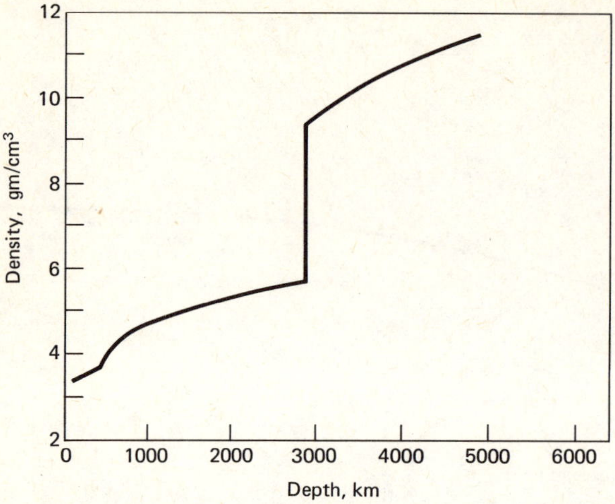

Figure 7.4 Variation of density with depth in the earth. After Bullen.

Once the density is known for each depth, the pressure is easily found. Consider the earth as divided into spherical shells of thickness δr. The pressure at any depth is found by adding up the weight of all the shells above. Mathematically, the procedure is as follows: The increment of pressure from top to bottom of each shell, δP, is

$$\delta P = \rho g \delta r \tag{7.1}$$

where ρ is the density of the material in the shell. The acceleration of gravity at each shell is

$$g = Gm/r^2 \tag{7.2}$$

where m is the mass contained in the sphere of radius r. From Eqs. 7.1 and 7.2,

$$\delta P = (Gm\rho/r^2)\delta r \tag{7.3}$$

Starting at the surface, the increment in pressure in each shell can be added to find P at any depth. The result is shown in Figure 7.5. To appreciate the magnitude of this pressure remember that 1 bar $= 10^6$ dyne/cm$^2 \simeq 1$ atmosphere. At the center of the earth the pressure is about 3.6 million atmospheres.

EARTH MATERIALS

To understand the motions within the solid earth it is necessary to know not only the temperature and pressure of the interior but also something of the materials which are present. The different species of materials found in the earth are called *minerals*. Quartz, calcite and rock salt are familiar examples. Almost all minerals have an internal structure which is *crystalline,* meaning that the constituent atoms are arranged according to a regular pattern. (More information on minerals is given in K. K. Turekian's *The Chemistry of the Earth.*) In the crust there occurs an extraordinary variety of minerals, many with complex chemical compositions and crystal struc-

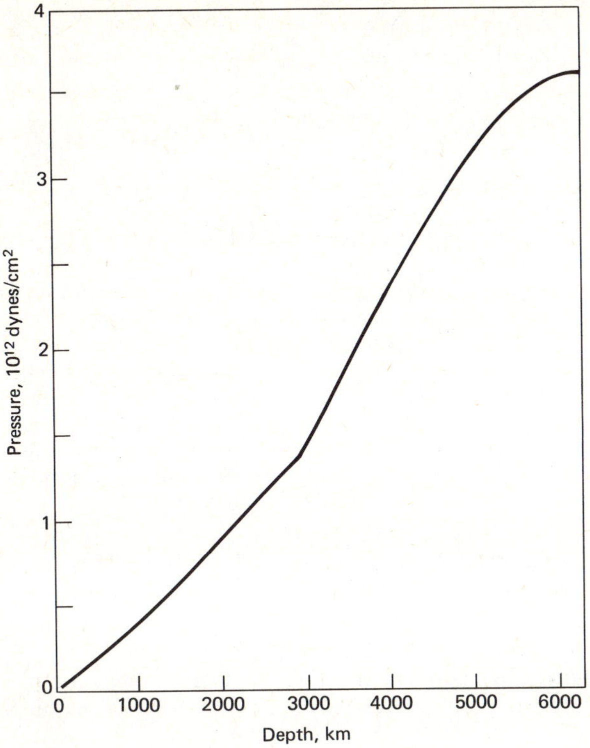

Figure 7.5 Variation of pressure in the interior of the earth. After Bullen.

tures. The mineralogy of the mantle (which accounts for 70 percent of the earth's mass) is simpler. Where samples of mantle material are exposed at the surface, as in the *diatremes* (volcanic vents) of southeastern Utah, the composition is principally magnesium silicate (Mg_2SiO_4) and, in lesser amounts, iron silicate (Fe_2SiO_4). In the crystal structure of these silicates, the oxygen atoms, which are relatively large, are packed together as closely as possible.

There are interstitial spaces between the oxygen atoms where smaller atoms such as Si and Mg can be inserted, as in Figure 7.6. However, the Si and Mg atoms are a bit too large to fit the interstitial spaces without some distortion of the oxygen packing. The mineral forsterite (Mg_2SiO_4), a constituent of the upper mantle, has a nearly close-packed array of oxygen atoms with the appropriate proportions of Mg

SILICATE STRUCTURES

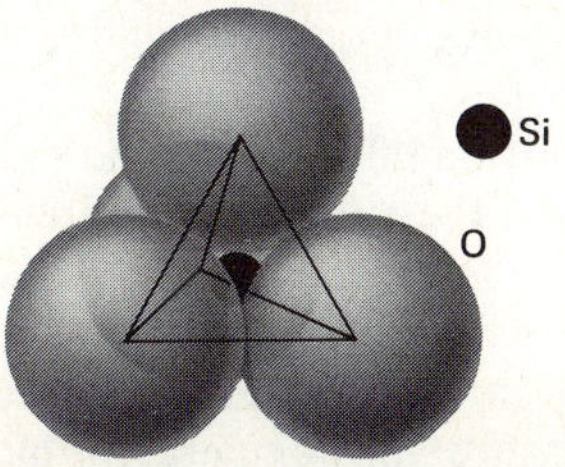

Figure 7.6 A silicon atom inserted in the interstitial space between nearly close-packed oxygen atoms.

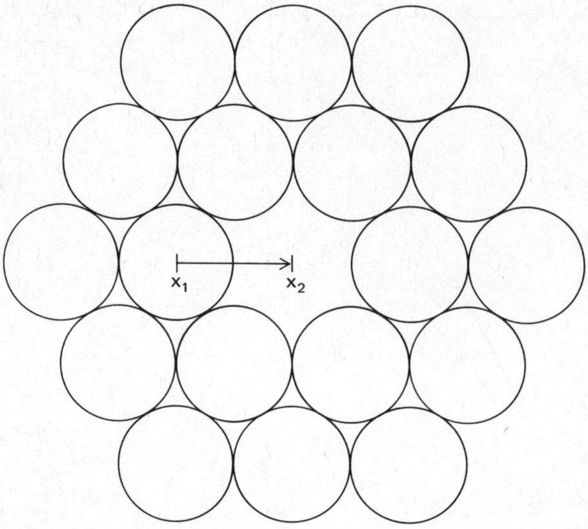

Figure 7.7 A vacancy exists at position x_2 in this close packed crystal. To the left the atom at x_1 is blocked by a near neighbor but to move to the right, into the vacancy, the atom at x_1 can squeeze between atoms above and below.

and Si atoms inserted in the interstices. Forsterite is stable at room pressure but at a sufficiently high pressure there is a shift of atomic positions to closer packing of the oxygen atoms. This is a *phase change:* At a critical pressure forsterite decomposes and is replaced by the higher density mineral *spinel;* at yet higher pressures spinel in turn decomposes into MgO (*periclase*) and *stishovite,* a high density form of silica. (The term "phase" as used here means a physically distinct state; it has nothing to do with a seismic "phase.")

At room temperature and pressure a crystal of forsterite is strong and brittle—under stress no deformation occurs until, when a sufficiently large stress is reached, the crystal shatters. However, the mantle is not rigid and unyielding. At high temperature, magnesium and iron silicates become plastic and can undergo large deformations without breaking. The ability of crystalline minerals to sustain large plastic flow at high temperature is a consequence of *atomic mobility*. This movement is much facilitated if some atom sites in the crystal are unoccupied (as they are in any real crystal). Figure 7.7 shows an unoccupied site (a *vacancy*) in a close-packed crystal. It is a relatively easy matter for an atom to squeeze past its neighbors and to move into the vacant space. The atom at x_1 might move to position x_2, for example, creating a vacancy at x_1. By successive jumps of this kind a complete rearrangement of the atoms in a crystal can be made. The greater the temperature the more vigorously atoms in a crystal vibrate. A high amplitude of vibration makes it much more likely that an atom will jump into a neighboring vacancy. This means that atomic mobility increases as the temperature increases. At high temperature the atomic jump rate becomes very large and atoms can move about through the crystal quite freely.

The application of hydrostatic pressure has a significant effect on atomic mobility. Squeezing the atoms of a crystal closer together makes the atomic interchange illustrated in Figure 7.7 more difficult. This is because, as the crystal is compressed, it is harder for an atom to squeeze past its neighbors to reach an adjoining vacancy. Thus application of pressure has the opposite effect of an increase in temperature; large scale plastic flow becomes more difficult. Both pressure and temperature increase with depth in the earth. The plasticity of the solid earth responds, then, to two opposing influences: The higher temperature at greater depth facilitates plastic deformation, the higher pressure opposes it. There is, however, an important difference in the form of the pressure and temperature profiles (Figs. 7.3 and 7.5). Temperature rises rapidly near the surface, then levels off; pressure increases regularly with depth. In the upper mantle, temperature is relatively high and, indeed, is near to the melting temperature of the rocks thought to be present, while the pressure is relatively low. In this zone near the top of the mantle the susceptibility of rock to plastic deformation is greatest; going either upwards or downwards, the resistance to deformation increases. This qualitative result is well established; making the statement quantitative remains a difficult problem. The actual temperature profile is imperfectly known, as is the exact composition of the rock of the mantle. Experimental observations which would indicate the depth dependence of the mantle plasticity would be of great value. Unfortunately, the available experimental evidence is scarce and not very satisfactory. Most of what is available comes from study of the earth's shape and from observation of the depression of the earth's surface under very heavy external loads.

THE EARTH'S FIGURE

If the earth were composed of material having no shear strength and were not rotating, its equilibrium shape, aside from the tidal bulges, would be a sphere (assuming also only radial variations in composition) because this would minimize the gravitational energy. Because the earth is rotating about its axis, centrifugal forces act radially outwards on material in the vicinity of the equator. If the earth had no internal shear strength, rotation would cause it to flatten; the equatorial radius would increase until the centrifugal force was balanced by gravitational force. This flattening is illustrated in Figure 7.8. The amount of flattening is measured by the ratio $f = (c - a)/a$. For an earth composed of material with no shear strength, theory shows that $F = 1/299.75$. It is now possible to measure f very accurately by study of the orbits of artificial satellites. The observed value of f is $1/298.25$. Thus the earth has very nearly the shape which would be expected if it were composed of material having no shear strength. This proves that plastic flow can occur in the solid mantle of the earth. However, the difference between the theoretical value of f for an earth with no shear strength and the measured value means that the mantle is sustaining a shearing stress of about 6 bars. This is a low stress, lower than the yield strength of the softest metals, for example.

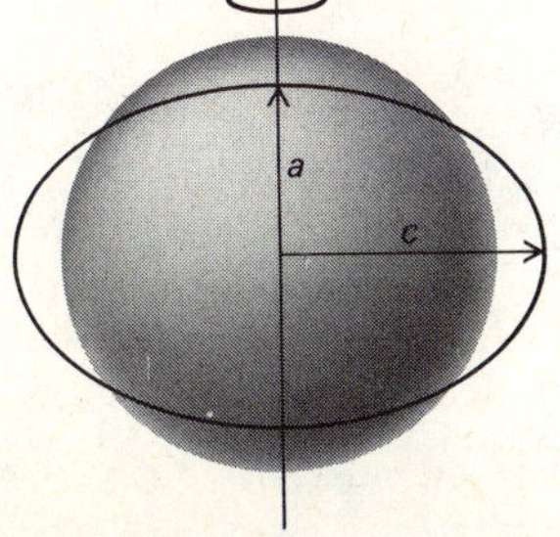

Figure 7.8 Rotation of a non-rigid earth results in flattening at the poles and expansion at the equator.

The observed flattening of the earth is *greater* than it would be if the upper mantle had no shear strength. The simplest interpretation of this observation is that at an earlier time the rotation rate of the earth was higher than it is now. We saw in Chapter 3 that the rotation rate of the earth is being slowed by tidal friction. The flattening of the earth should become smaller as the rotation rate decreases. At the present rate of deceleration the time required to change the flattening of an earth lacking shear strength by an amount as great as the present discrepancy in f is about two million years. On the geologic time scale, where the oldest surface rocks date to 3000 million years, this is a relatively short period of time. Tidal deceleration at the present rate has surely been active for more than two million years. Two interpretations of these observations are possible. Either the mantle has an overall strength of about 6 bars which must be exceeded before any deformation is possible, or the mantle deforms at a rate slow enough so that there is always some delay in its response to the stresses produced by deceleration. Since the mantle is sufficiently hot throughout for high temperature flow to be active, it is most likely that the discrepancy in the flattening is due to delayed response of highly viscous material within the mantle rather than permanent strength. Since the strength properties of the mantle vary considerably with depth, and the depth at which the nonequilibrium figure is supported cannot be found, a quantitative interpretation of the flattening discrepancy is hardly possible.

ISOSTATIC COMPENSATION

A second source of information about the strength properties of the mantle comes from observations of the response of the surface of the earth to the application of large forces of long duration. It is difficult to generate forces on the surface large enough to produce a deformation extending through the crust and into the mantle. Fortunately, nature has done the experiment for us. During the last ice age the land of Canada and Scandinavia was burdened with the weight of ice sheets up to 2500 m thick. In Figure 7.9a, a layer of crustal rock is shown over the denser mantle rock below. If a mass of ice is placed on the crust, as in Figure 7.9b, the crust will be forced down deeper into the mantle below. The weight of the ice is supported by the increased buoyant force resulting from the greater submergence of the low density continental crust. (This is analogous to the increase in depth to which a ship floats as it is laden with cargo. Buoyant support of a surface load on the earth is called *isostatic compensation.*) Under the thickest ice sheets of the last glacial age the subsidence of the crust would have been a thousand meters or more. Crust depressed by ice sheets exists today in Greenland and Antarctica. The depression of the central land of Greenland is shown in the cross section of Figure 7.10. In Antarctica only the tops of a few high mountains, called *nunataks,* penetrate the ice sheets.

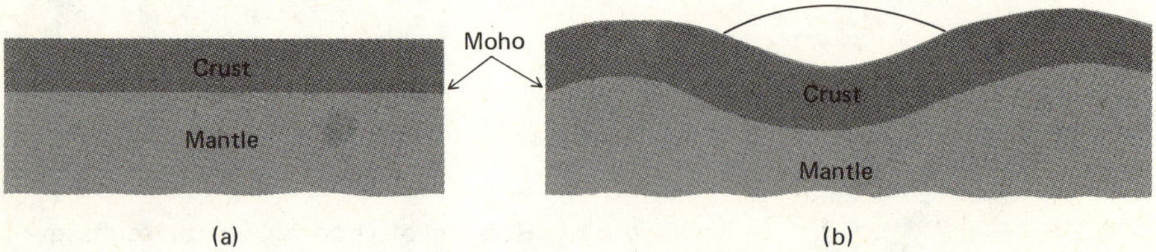

Figure 7.9 A layer of continental crust burdened with a load of ice must sink deeper into the mantle below in order to support the additional weight.

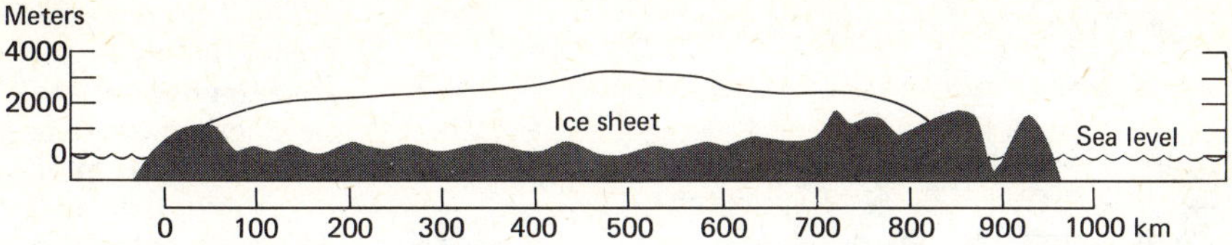

Figure 7.10 Cross section of Greenland showing the depression of the central land under the load of ice above.

The last ice age ended suddenly about 10,000 years ago; the ice sheets covering Canada and Scandinavia melted and the surface load was removed. The land, relieved of its ice load, began to rise in response to the now excessive buoyant forces below. It is still rising today because isostatic balance has yet to be fully recovered. This rise is recorded by geologic evidence and can be observed directly. For example, new islands have appeared in Hudson's Bay within the memory of living Eskimos. Study of the rate of uplift should yield evidence about the rate of plastic flow in the upper layers of the mantle where isostatic compensation is accomplished.

A relatively simple example of the action of isostatic compensation is ancient Lake Bonneville. During glacial times, when rainfall was abundant, a large valley in what is now Utah was filled with water to a depth of about 300 meters to form Lake Bonneville. As a result of a change to drier climate at the end of the ice ages the lake lost most of its water; Salt Lake is its remnant. There is abundant evidence of the former high water level in Lake Bonneville: Waves on the lake cut beaches which remain visible today (Fig. 7.11). By tracing these beaches the outline of the ancient lake may be located. The interesting thing about these old shorelines is that they are

Figure 7.11 Remnant shorelines of ancient Lake Bonneville. Courtesy Ward's National Science Establishment.

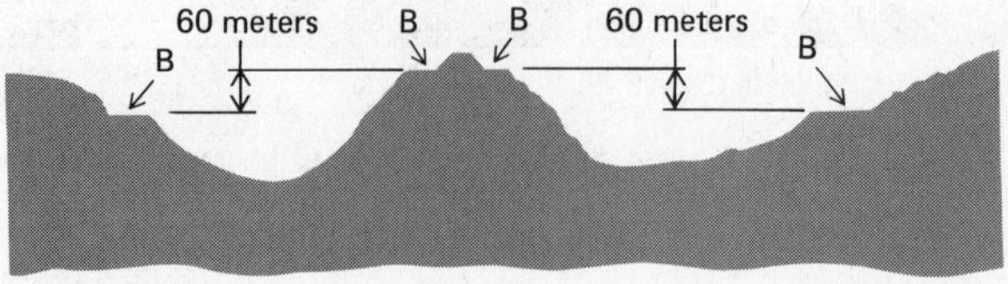

Figure 7.12 Schematic cross section through the Lake Bonneville basin. Beaches cut by waves in the ancient lake are marked B. Uplift of the lake bottom due to removal of the water load has raised the beaches on the former island by about 60 meters.

no longer level. This is illustrated by the cross section shown in Figure 7.12, where the shoreline at the outer edge and on a former island in the center of the lake is shown. The shoreline of the island now stands about 60 meters higher than the shoreline at the edge of the lake. This is evidently due to uplift of the lake floor in response to the removal of the water load. That uplift is continuing today is revealed when surveys made for the construction of the Union Pacific Railroad are compared with modern surveys: Where the railroad crosses the ancient lake bed it is higher now than when built.

Uplift of a section of the crust from which ice or water load has been removed continues over a period of many years because the necessary deformation of the underlying mantle is accomplished slowly. A quantitative analysis of the deformation rate under a surface load (or upon removal of the load) is possible, so far, only if the deforming material is assumed to be viscous. For this reason it is often assumed that the material of the upper mantle can be described as viscous; that is, its flow rate is proportional to the applied stress. (Independent justification of this assumption is not possible yet.) The more viscous the material, the longer the time required for the recovery process. The viscosity of the upper mantle found from the application of

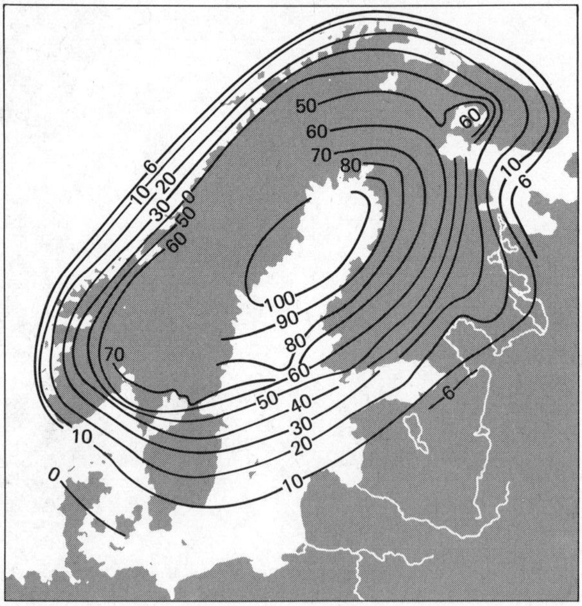

Figure 7.13 Present elevation (in meters) of shorelines in Scandinavia which were level at about 3000 B.C. Uplift centered near the head of the Gulf of Bothnia is evident. After Sauramo.

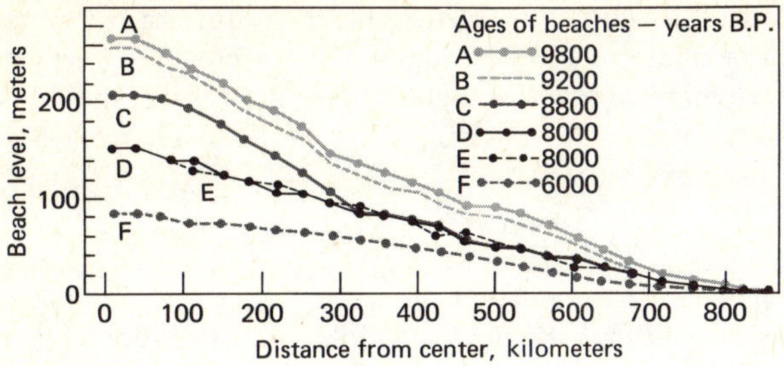

Figure 7.14 Present elevations of beaches of different ages measured outwards from the center of uplift. The time when each beach was cut is indicated. After R. K. McConnell, Jr., *Journal of Geophysical Research*, **73,** 7089 (1968).

this relation to the observed uplift at Lake Bonneville is of a magnitude (10^{21} poise) that can be easily accounted for by the atomic mobility in silicate minerals described in the preceding section.

Uplift on a much larger scale than in Utah is occurring in Scandinavia as a result of deglaciation. The greatest increase in elevation is near the head of the Gulf of Bothnia, where the ice sheet was thickest. This is shown in the map in Figure 7.13 where the present elevation of shorelines which were level about 5,000 years ago are plotted. The profiles of a series of once-level beaches along a line radiating out from the center of uplift are shown in Figure 7.14. The rapidity of the early stages of the uplift is clearly seen when the beach profiles at successive times are compared. Most of the uplift occurred between 6000 and 9800 years ago. Analysis of these data by McConnell shows that the uplift in the central region is faster than would be expected from a model based on the recovery of a surface indentation on a slab of material of uniform viscosity. This may be interpreted as evidence that the viscosity of the mantle is not uniform with depth. The best fit to the data is obtained when a model like that in Figure 7.15 is assumed: An elastic layer lies over viscous material in which the viscosity becomes larger as the depth increases. This is a further confirmation of the idea that there is a zone of minimum strength in the upper mantle. The results, however, cannot be applied directly to deformation of the mantle in tectonic processes: Isostatic compensation involves only very small strains and its linear equations cannot be extended to large-scale deformation.

PLASTIC PROPERTIES OF THE MANTLE

Theory indicates, and field observations clearly show, that plastic deformation occurs in the earth's mantle. On the basis of laboratory experience and theory, possible mechanisms of deformation can be identified and the general character of the depth dependence of the plasticity determined. Detailed prediction of the mechanical prop-

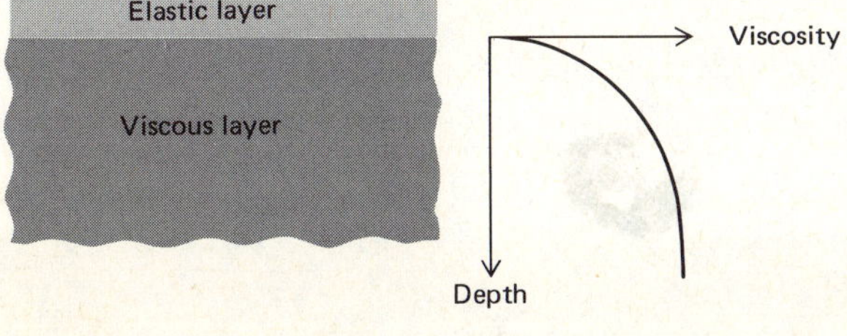

Figure 7.15 Earth model which best fits the Scandinavian upilft data. Under a surface layer of purely elastic material is viscous material whose viscosity increases with depth. After McConnell.

erties is not possible because the temperature gradient, the nature of the materials, and their chemical environment are not well enough known. Relevant observational data are sparse and there are many difficulties in their interpretation. It should not be surprising that it is difficult to identify the mechanisms of the large-scale earth movements to be discussed in the next chapter.

Further Readings

CLARK, S. P., *The Structure of the Earth.* Englewood Cliffs, N.J.: Prentice-Hall, 1971.

GIRIFALCO, L. A., *Atomic Migration in Crystals.* Waltham, Mass.: Blaisdell, 1964.

HAYDEN, H. W., et al, *The Structure and Properties of Materials, Vol. III, Mechanical Behavior.* New York: Wiley, 1964.

MOFFATT, W. G., et al, *The Structure and Properties of Materials, Vol. I, Structures.* New York: Wiley, 1964.

TUREKIAN, K. K., *The Chemistry of the Earth.* New York: Holt, Rinehart and Winston, 1972.

8 Motion of the Continents and the Formation of Mountains

According to the theory of isostasy, the buoyant force derived from less dense crustal rock floating on denser mantle rock supports the weight of mountains on the continents; the crust is thicker—the Moho deeper—under mountains. This view of earth structure describes a soft, static mantle on which the continents float much as icebergs float upon the sea. But there are active heat sources in the mantle. And heating of the mantle can set it into motion. Locally heated rock will tend to move upwards while relatively cooler rock sinks. Local heating can initiate large-scale, convective flow of the mantle, flow which could move the continents about over the earth's surface and supply the forces which cause earthquakes and the formation of mountains. Earthquakes and the folding and faulting of rocks are evidence of the action of earth-moving forces of large magnitude, but they give little indication of the world-wide force pattern. The discovery of sea floor spreading has made possible the determination of the overall pattern of motion of the earth's crust.

SEA FLOOR SPREADING

Chapter 2 showed that the geography of the ocean basins tends to be more regular than that of the continents. The world's largest mountain system is the long chain of mid-ocean ridge charted in Figure 2.16. The mountains of the mid-ocean ridge have a characteristic symmetry about a central notch and are broken into a series of blocks offset laterally as shown in Figure 8.1. These features are not found in continental mountains. The ocean basins are also marked by trenches, narrow valleys deeper than the highest mountain ranges on the continents. Two sets of observations suggest how the ridges and trenches have been formed: The data in Table 7.1 show that the heat flow near the ridges is higher, and that near the trenches lower, than the mean for ocean basins. Second, the deep sea drilling discussed in Chapter 2 shows that the sediment cover on the ocean floor near the ridges is very thin, vanishing on the ridge flanks (which are composed of the volcanic rock *basalt*), and becoming progressively thicker away from the ridges. Outwards from the ridges the bottommost layer of sediment becomes progressively older, but even the oldest sediment found on the sea

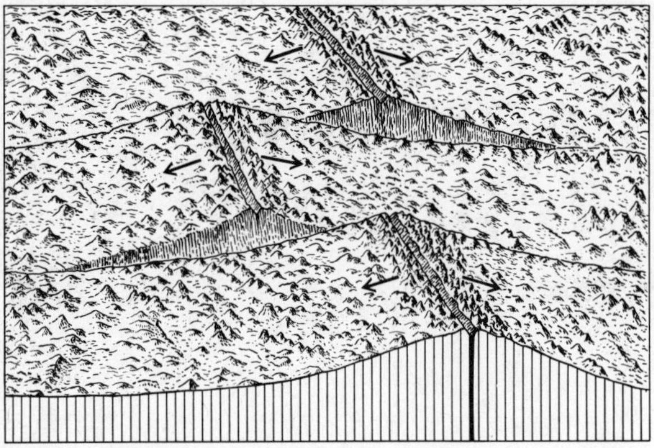

Figure 8.1 The mid-ocean ridges are marked by a deep central notch and by numerous offsets at *transform faults.*

floor is younger than most of the continental rocks. All these observations suggest that new sea floor is created at the mid-ocean ridges by the upwards motion of hot rock from deep in the earth, that the rock brought to the surface at the ridges moves outwards across the ocean basins accumulating a growing cover of sediments as it moves, and that old sea floor is consumed by downward flow into the mantle at the ocean trenches. This is the hypothesis of sea floor spreading. The upwards flow of volcanic rock at the ridges (where the heat flow is high), its motion outwards and then downwards into the trenches is shown schematically in Figure 8.2. The hypothesis of sea floor spreading is beautifully confirmed, and its rate revealed, by the pattern of magnetization of the ocean floor.

MAGNETIC ANOMALIES OF THE SEA FLOOR

Volcanic rock contains among its constituents small amounts of ferrimagnetic minerals, chiefly magnetite. A *ferrimagnetic* mineral can be permanently magnetized; magnetite, or lodestone, was the first permanently magnetized material to be discovered. Permanent magnetization of a material is possible only below a certain characteristic temperature called the *Curie temperature*. If, for example, a piece of iron is heated, its permanent magnetization weakens, slowly at first, then more rapidly, until

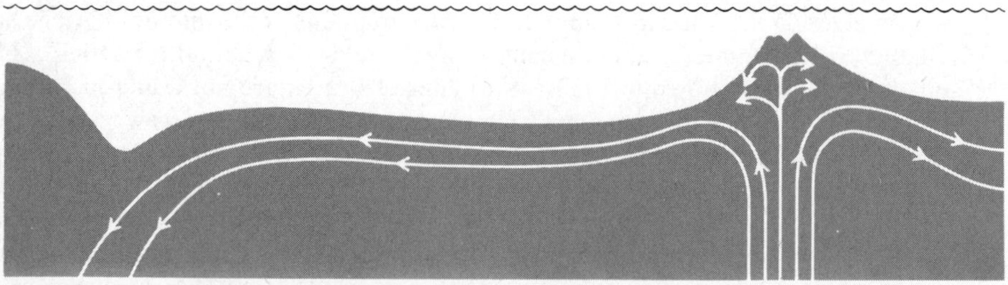

Figure 8.2 The ocean ridge and trench structure is interpreted as the surface manifestation of a large scale pattern of flow in the mantle below. Under the ridges mantle material rises and spreads outward; trenches are located over downward flow. Sediment cover thickens away from the ridge.

at the Curie temperature it disappears. The process is reversible and the magnetization reappears upon cooling of the iron. When hot volcanic rock is brought to the surface and cooled through the Curie temperature of its constituent minerals, it acquires a small but permanent magnetization in the direction of the earth's magnetic field.

The magnetic field of the earth is, to a good first approximation, the same as the field of a giant bar magnet (a magnetic *dipole*) placed at the earth's center (Fig. 8.3). The axis of the dipole does not quite coincide with the axis of rotation of the earth; the magnetic north pole is in northern Canada rather than at the geographic north pole. This means that a magnetic compass does not point true north in most parts of the world. Indeed, near the geographic north pole the "north" end of a compass needle will point nearly south. The difference in direction between magnetic north and geographic north at any place is called the *magnetic variation* at that place. Its magnitude can be found by comparing the direction of a compass needle with the direction of true north as determined by astronomical observations. Repeated observations over a period of years show that the magnetic variation at a given place changes slowly. The earth's magnetism is not constant in direction. Nor is it constant in magnitude. Since the time of the first accurate measurements of the intensity of the earth's magnetization early in the nineteenth century, the field strength has decreased by about 5 percent.

Although changes are revealed, the time span of instrumental observations is not long enough to show much about the pattern of alteration of the earth's field. A much longer record of the strength and direction of the magnetic field of the earth can be read from the permanent magnetization of volcanic rock which has remained undisturbed since the time it cooled through its Curie temperature. The magnetization of such rock measured in the laboratory today shows the direction and intensity of the earth's field at the time that the rock was cooled. Many examples of magnetized volcanic rock of known age (see K. K. Turekian's *Chemistry of the Earth* for a description of how rock ages are determined) are found around the world; some of the best are the lava flows of the Hawaiian volcanoes. The magnetization of this rock shows that changes in the magnetic variation are due mainly to small, variable irregularities

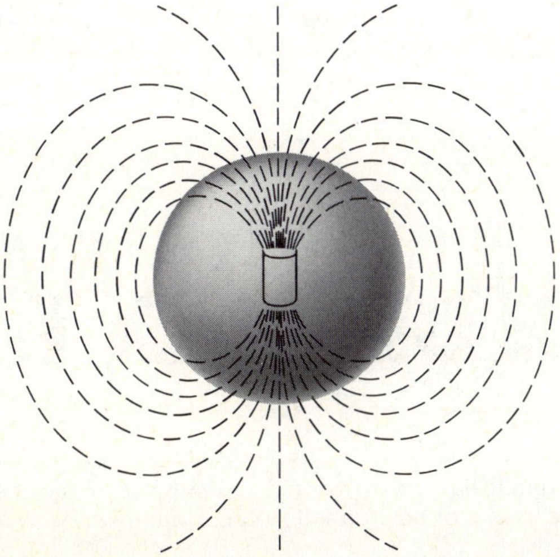

Figure 8.3 The earth's magnetic field is approximately the same as that of a magnetic dipole (bar magnet) located within at the center but inclined somewhat to the axis of rotation. (After Bowditch.)

in the dipole field, that the earth's magnetic dipole remains nearly in the north-south orientation, but that the direction of the magnetization reverses quite abruptly from time to time. During a "normal" period magnetic north is near geographic north; during a "reversed" period magnetic south is near geographic north. The observed sequence of field reversals is not regular; the field remains in one direction for ten thousand to perhaps a million years, then suddenly reverses. The observed sequence of reversals is shown in Fig. 8.4.

What causes these reversals remains a mystery. The earth's magnetic field is thought to arise from the motion of the liquid part of the core. The liquid metal acts like the moving conductors of a dynamo to generate electric currents which are responsible for the magnetic field. As the pattern of motion in the core changes, the magnetic field at the earth's surface will also change. That the direction of the magnetic field should reverse from time to time is a reasonable possibility, since the flow pattern in the outer core may not be stable, but a detailed theory has not yet been constructed.

Assume that volcanic rock is brought to the surface at the mid-ocean ridges and that it then moves outwards, cooling as it moves. A short distance from the ridge it falls below its Curie temperature and becomes permanently magnetized in the direction which the earth's field has at that time. Rock brought to the surface at a later time may cool through its Curie temperature when the earth's magnetism is reversed. Reversals in the direction of the earth's field will be recorded in reversals in the magnetization of the volcanic rock spread along the sea floors; successively older reversals are found further away from the mid-ocean ridge. Reversals in the permanent magnetization of the rock on the sea floor cause small changes (anomalies) in the strength of the magnetic field measured at the surface. These magnetic anomalies can be detected with a *magnetometer,* an instrument which measures magnetic intensity very accurately. (Sensitive magnetometers were first developed to detect submerged submarines by sensing the magnetic anomaly caused by the iron hull of the submarine.) A magnetometer moving outwards from any active ocean ridge should then show a characteristic sequence of anomalies in the earth's magnetic field. In fact, the observed anomaly patterns can be matched with the time sequence of field reversals. This is illustrated in Figure 8.4.

The time at which each field reversal occurred is known (Fig. 8.5). If the distance of a particular reversal from the ocean ridge source is measured, then the speed of

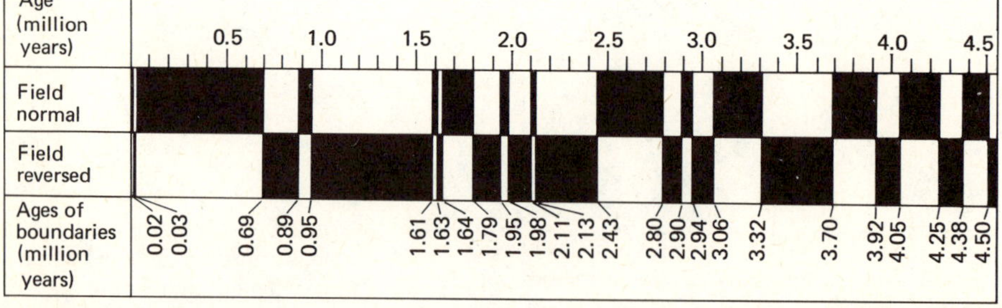

Figure 8.4 The direction of the earth's magnetic field has reversed frequently during the past 4.5 million years. On this time scale, reversals of field direction occur quite suddenly. After Cox, *Science*, **163,** 237 (1969).

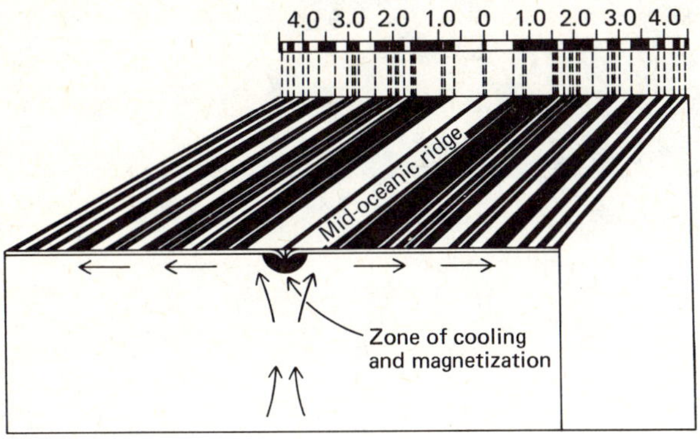

Figure 8.5 Shown here schematically is the coincidence of magnetic anomalies observed on the sea floor and the sequence of reversals of the earth's magnetic field. The coincidence is interpreted in terms of sea floor spreading; the spreading rate can be determined from the distance of the anomalies from the axis of spreading. (After Cox, Dalrymple and Doell.)

sea-floor spreading is fixed. Spreading rates measured in this way range up to 10 cm/yr. They show an outwards movement of sea floor away from the mid-ocean ridges and towards the ocean trenches. This is a quantitative confirmation of the hypothesis of sea-floor spreading.

PLATE TECTONICS

Comparison of the plotted positions of earthquake epicenters (Fig. 6.22) with the ocean floor chart (Fig. 2.16) shows a remarkable coincidence of epicenters with the ocean ridges and trenches. This coincidence suggests that the tectonic movements responsible for earthquakes are the same as those revealed by the spreading of the sea floor. This hypothesis is confirmed by the fault motion directions of earthquakes determined on a world-wide basis. Some of the fault motion determinations which have been made are shown in Figure 8.6. They clearly indicate spreading of sea floor outwards from the mid-ocean ridges.

The charted earthquake epicenters of Figure 6.22 and the fault motion directions of Figure 8.6 show that the surface of the earth may be divided into a set of areas within each of which all surface material is moving uniformly in the same direction. These areas have been called *plates* to indicate the fact that they move as if they were rigid sections of the earth's crust. (The term "plate" is somewhat misleading because the apparent plate rigidity is a manifestation of the uniform underlying motion in the mantle, not of the structural integrity of the "plates.") The system of plates indicated by seismic and magnetic evidence is charted in Figure 8.7. These plates are in motion relative to each other. At a plate boundary marked by a mid-ocean ridge the plates are moving apart, the gap so created being filled by new oceanic crust made from hot mantle rock brought up from below. Where two plates moving toward each other abut under the ocean, oceanic crust is absorbed into the mantle (since there is no evidence for significant expansion of the earth). Such a junction is marked by a trench, as the Tonga Trench at the abutment of the Indian and Pacific plates (Fig. 8.7). When the boundary of oppositely moving plates is parallel to the direction of motion, the boundary is called a *transform fault*. A transform fault and trench are illustrated in Figure 8.8.

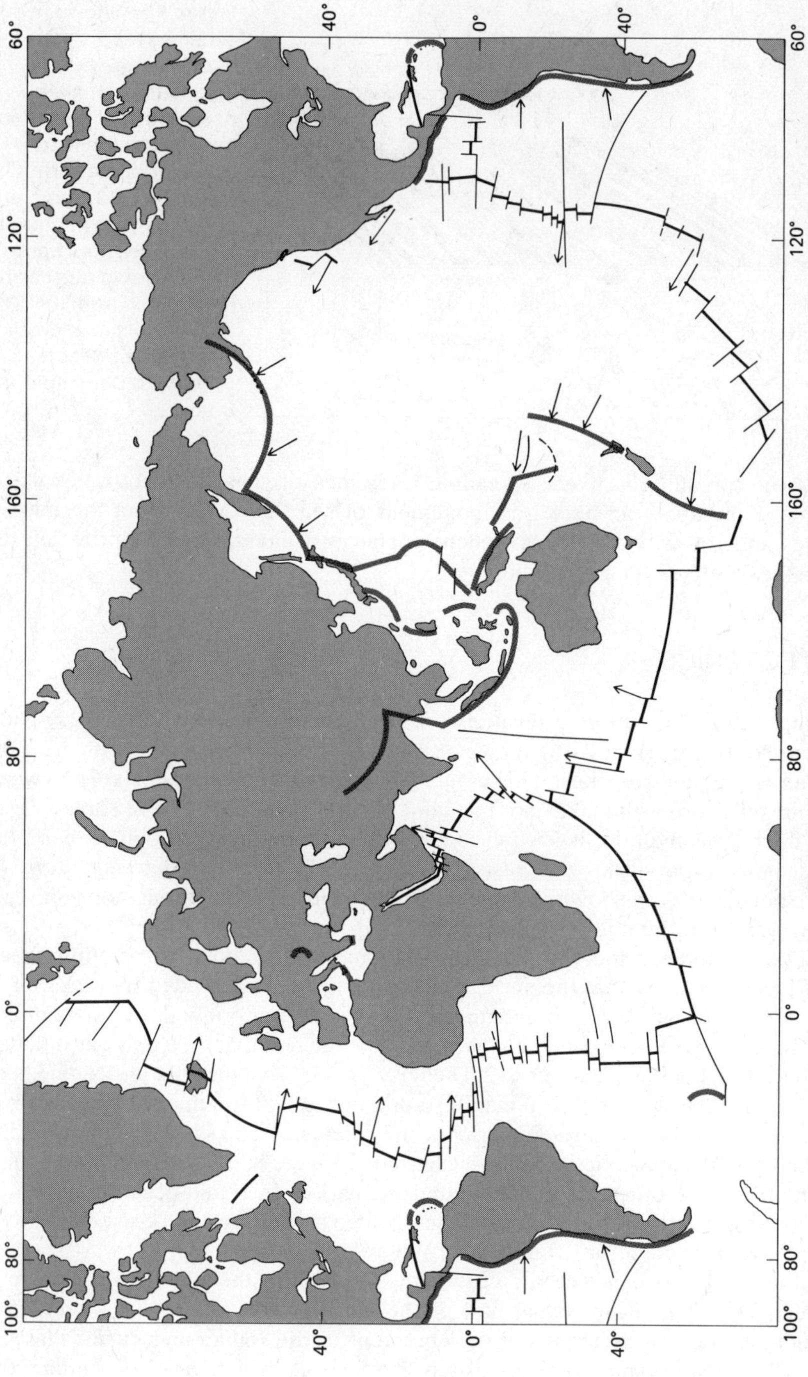

Figure 8.6 Arrows show relative displacements at the faults responsible for observed earthquakes. The pattern of motion suggests that the earth's surface may be considered to be divided into a set of "plates" which are in motion relative to one another. Based on the data of Isacks, Oliver and Sykes, Journal of Geophysical Research, **73,** 5861 (1968).

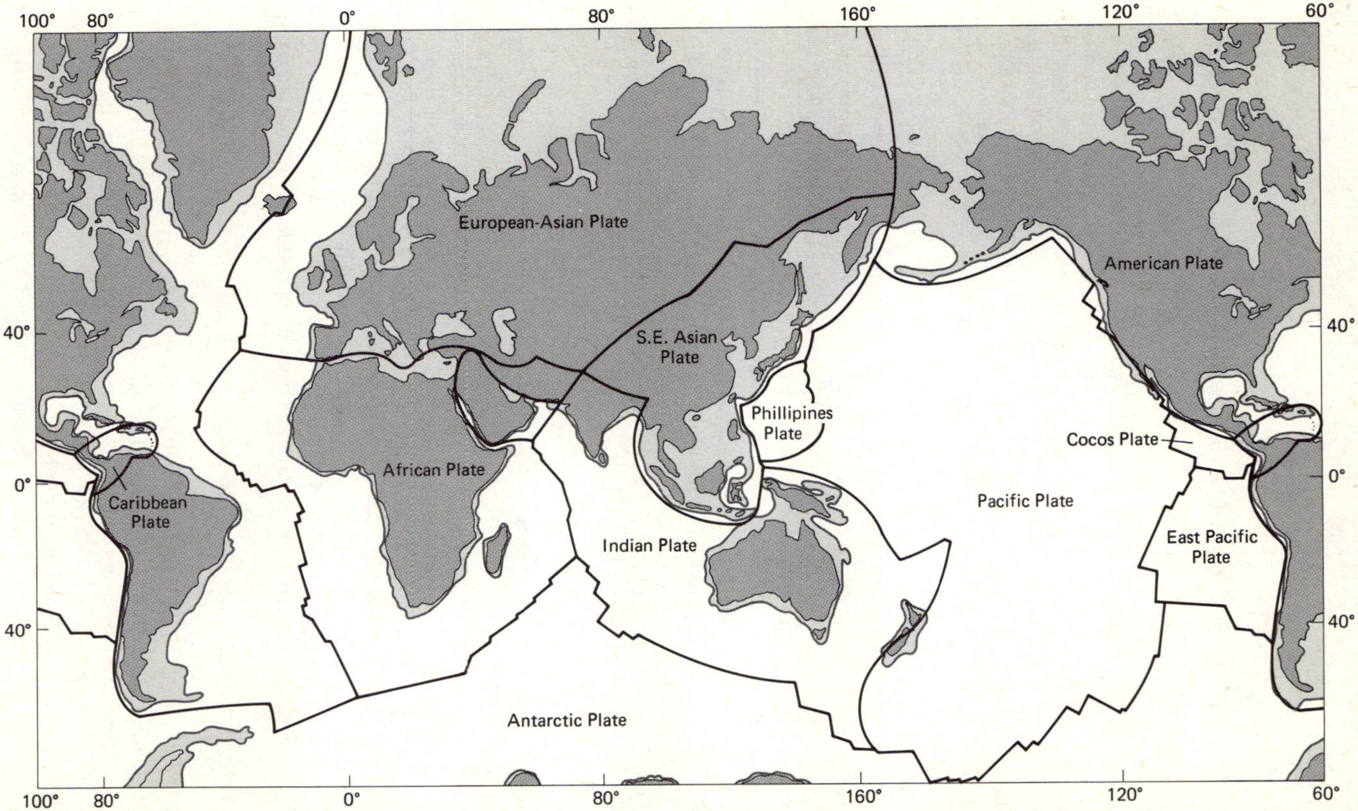

Figure 8.7 The world-wide system of tectonic plates constructed from observations of ocean bottom topography, and sea floor spreading.

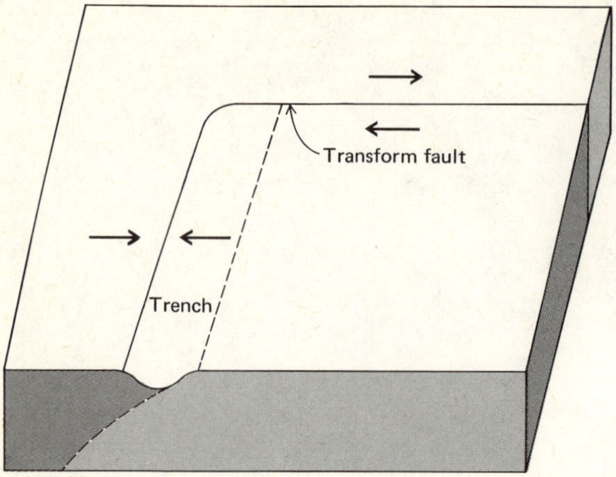

Figure 8.8 Where the relative motion of two plates is convergence from opposite directions, the meeting point may be marked by a trench. Where the motion is parallel to the plate boundary, a transform fault is formed.

Continents move along with the plate on which they happen to be placed. If a plate boundary runs under a continent then the two parts of the continent will be in relative motion. An example is the San Andreas fault in southern California: It is a transform fault which results from the northwards motion of the Pacific plate relative to the American plate, which underlies most of continental North America. Earthquakes are frequent along the San Andreas fault; the east coasts of North and South America are generally free of earthquakes because they are far removed from any plate boundaries.

To further study the motion of plates, it is helpful to study in more detail the structure of ocean ridges and trenches.

STRUCTURE OF THE RIDGES AND TRENCHES

The mid-ocean ridge profile, with its central notch, is suggestive of a surface feature formed by the upwards and outwards motion of the material below, as in Figure 8.2. Many earthquakes occur along the mid-ocean ridges; they are found to be always shallow focus earthquakes. This is to be expected since the material rising at the ridges is, as it comes from the upper mantle, relatively hot; it only becomes cool enough to begin fracturing when it comes quite close to the surface. The sides of the ridges are steep, free of sediment, and covered with blocks of the volcanic rock basalt.

A structural feature of the ridges not so easily explained is the off-sets of the ridge axis at the transverse fracture zones illustrated in Figure 8.1. Determination of the relative motion of these fracture zones can be used, however, to provide another test of the sea-floor spreading hypothesis. Figure 8.9 shows the relative motions to be expected along the fracture zones in the case (a) where spreading of the sea floor outwards from the ridge axis is the only cause of motion on the fracture and (b) where fracture is actively displacing the ridge axis and there is no spreading. In the former case earthquake activity should be largely confined to the space between the two ridges, since outside of this zone the crust is moving with the same speed and direction on both sides of the fracture. In fact, the observed epicenters are so confined.

The ocean trenches, the places where ocean floor is consumed, also have a char-

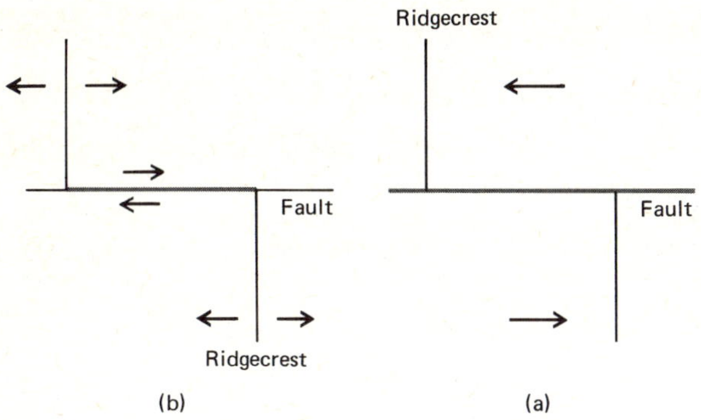

Figure 8.9 (a) Where an ocean ridge is offset and spreading from the ridges is active, slippage on the fault is confined to the zone between the two ridge segments. But if the fracture zone is interpreted as a fault actively offsetting the ridge's axis, as in (b), then slippage would occur over the entire fault length. The fact that earthquake epicenters on the fracture zone are observed only between the ridge axes is a further confirmation of sea-floor spreading.

acteristic structure. The Tonga Trench lying to the east of the Friendly Islands is a good example. The plan and cross section are shown in Fig. 8.10. The three parallel, linear features are the trench, which ranges to 9000 meters deep, a band of coral fringed islands, and a second band of islands which are sharp, volcanic peaks. (Tofua is a steep-sided former volcano with a lake in the old crater at the summit; it was the starting point of Captain Bligh's epic open-boat voyage. Metis Shoal is a new island just emerging from the sea.) Beneath the Friendly Islands there is intense, deep focus earthquake activity. This seismicity has been studied in detail by Sykes, Isacks, and Oliver. The earthquakes originate in a band only 20 km thick sloping downwards under the islands. There are marked directional differences in the propagation characteristics of seismic waves originating from these earthquakes. These are summarized in Figure 8.11. There is a zone of low attenuation which extends along the ocean floor eastward of the trench and which then dips down under the islands following the same downward trend as the zone of seismic activity. (Attenuation, α, is the decrease in wave amplitude with increasing distance traveled.) Seismic signals propagated outside this band, as are the signals to stations in the Fiji Islands, have high attenuation. The band of low attenuation is identified with the westward-moving Pacific plate; it must be, then, that under the Friendly Islands the plate is turned downward into the mantle below. The plate material moving downward from the surface is relatively colder than the surrounding mantle material. Low temperature implies low atomic mobility and the absence of fluid phases. These are the conditions favorable to low attenuation of stress waves. This is direct physical evidence that trenches are sinks for the crust which must disappear when plates colide.

The downwards moving plate material contains mineral species which become unstable at depths of hundreds of kilometers. Phase changes result as crustal materials are carried down into the mantle and may be one source of deep seismic activity. The plate as it moves down under the trench evidently carries some of the overlying, consolidated sediment with it. This consolidated sediment is rich in hydrous minerals (minerals incorporating water in their structures). As these minerals are exposed to the increasing temperature and pressure at depth they begin to decompose, releasing volatile constituents and, eventually, liquid formed by partial melting. The fluid and

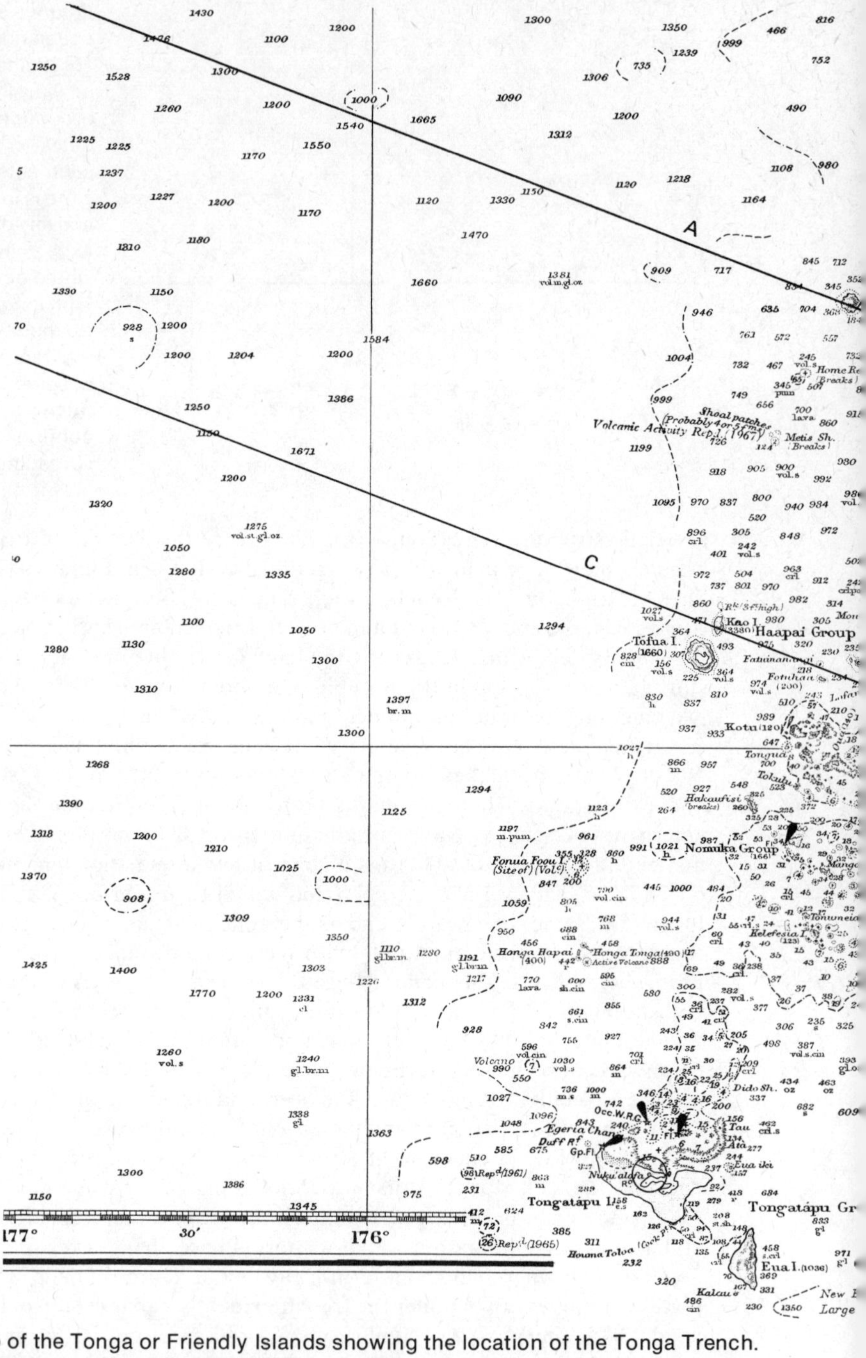

Figure 8.10(a) Map of the Tonga or Friendly Islands showing the location of the Tonga Trench. Depths are in fathoms; the 1000, 2000, 3000, and 4000 contour lines are shown with water depths indicated by shading. Reproduced from BA chart number 1829 with the sanction of the Controller, HM Stationery Office, and of the Hydrographer of the Navy.

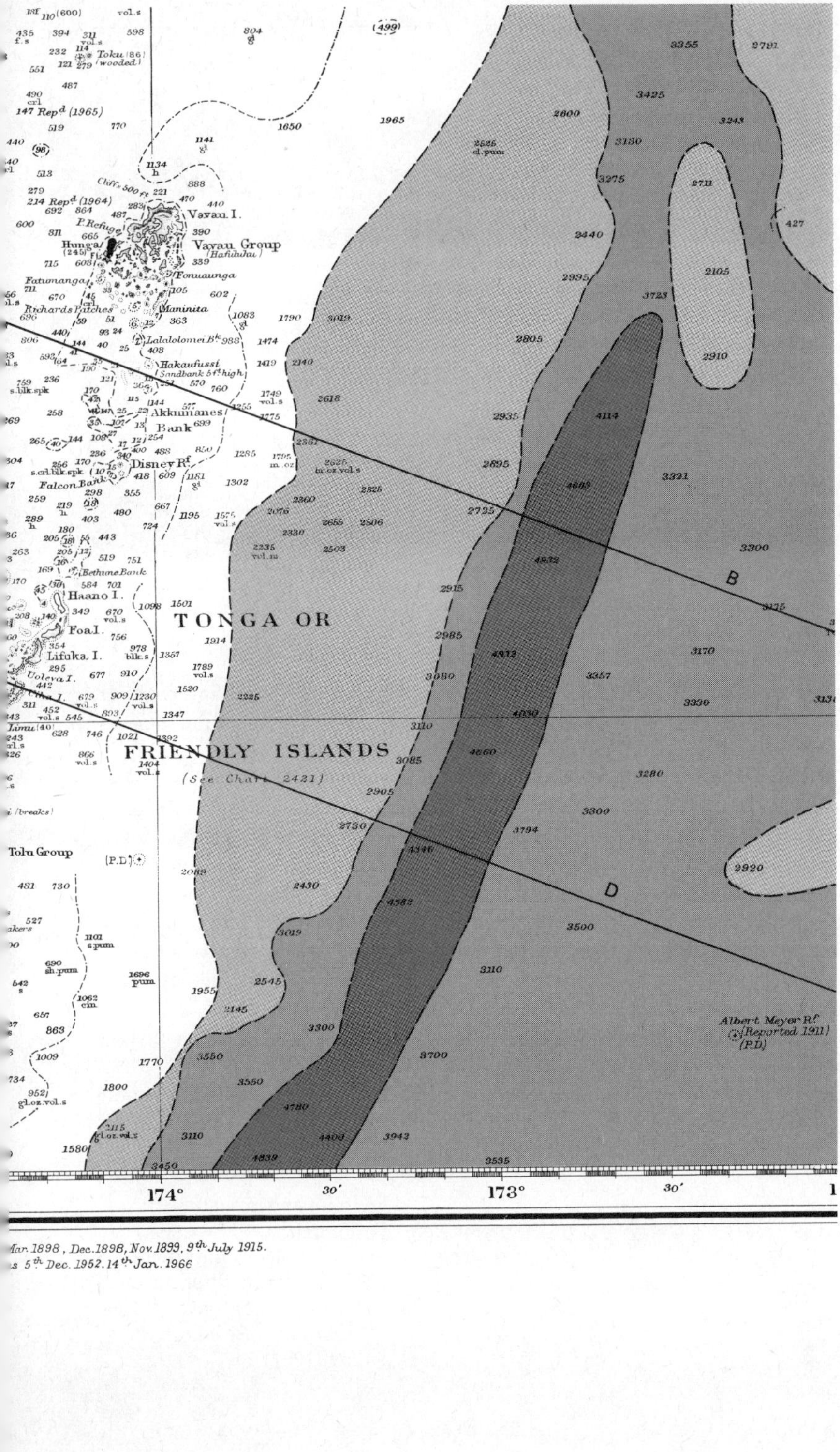
TONGA OR
FRIENDLY ISLANDS
(See Chart 2421)
Toku (86) (wooded)
Vavau I.
Vavau Group
(Hafulu Hou)
Hunga
P. Refuge
Fonuaunga
Fatumanga
Richards Patches
Maninita
Lalalolomei Bk.
Hakaufussi
Sandbank 5 ft high
Akkunanes Bank
Disney Rf.
Falcon Bank
Bethune Bank
Haano I.
Foa I.
Lifuka I.
Uoleva I.
Tofua Group
(P.D.)
Albert Meyer Rf.
(Reported 1911)
(P.D.)
B
D
174°
30′
173°
30′
Mar. 1898, Dec. 1898, Nov. 1899, 9th July 1915.
5th Dec. 1952. 14th Jan. 1966

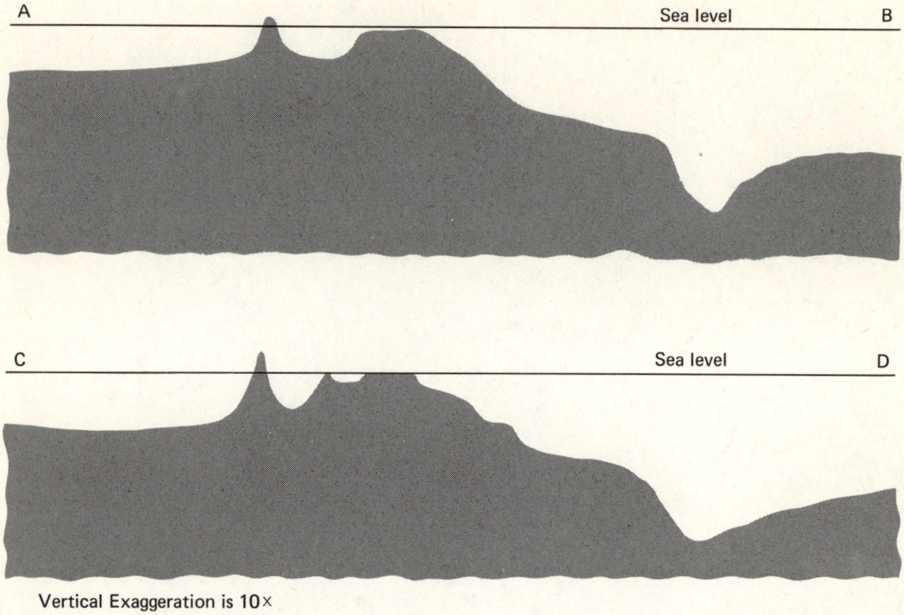

Figure 8.10(b) Cross sections of the island and trench structure along the lines AB and CD shown in (a).

volatile material so released makes its way upwards through the overlying rock. It creates a zone of high seismic attenuation (as in Fig. 8.11), and upon reaching the surface, escapes through the volcanoes of the second arc of islands.

The rate of descent of plate material must match the rate of sea floor spreading towards the trench. The depth to which the plate can penetrate the mantle is fixed by the time it takes to become heated to the point where plate materials are no longer stable; when that temperature is exceeded the plate has decomposed. Heating of the downwards moving plate is slow because of the small thermal conductivity of rock. It follows that the faster the rate of sea floor spreading, the greater the rate of descent of the plate under the arc, and the greater the depth that plate material will attain before it reaches the decomposition temperature. Observations which support this interpretation are shown in Figure 8.12.

The seismic attenuation observations indicate a plate thickness of about 90 km under the ocean floor. The plates are in continuous motion but it is unlikely that they are being pulled or pushed along the sea floor. It is unlikely that they are being pulled because the tensile strength of rock is characteristically low. (This is why a Gothic

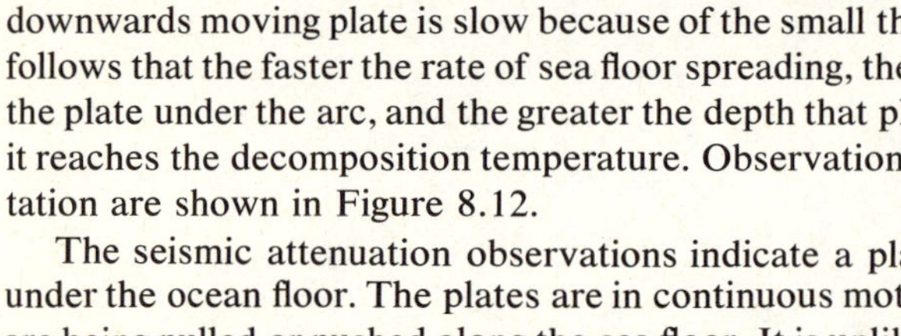

Figure 8.11 Zones of high (white) and low (shaded) attenuation, α, for seismic waves in the vicinity of the Tonga Trench. After Oliver and Isacks, *Journal of Geophysical Research,* **72,** 4259 (1967).

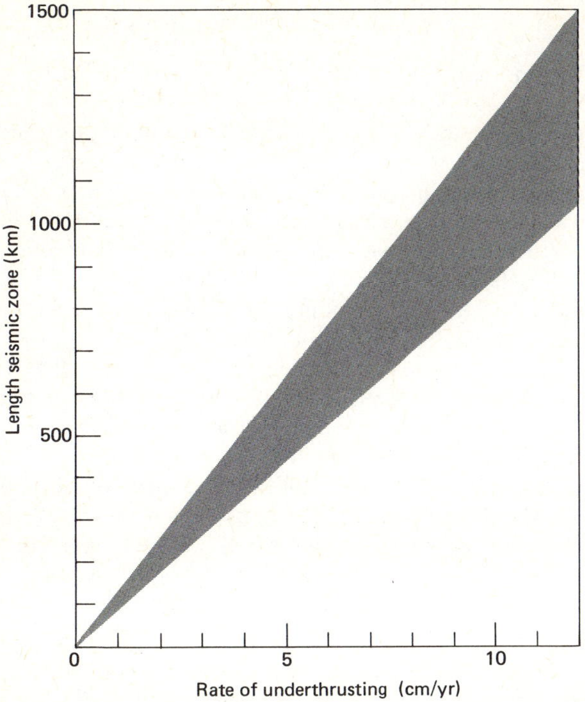

Figure 8.12 The greater the rate at which two plates move together, the greater the length of the seismic zone at a trench. After Isacks, Oliver and Sykes, Journal of Geophysical Research, **73,** 5885 (1968).

cathedral or other stone structure is designed so that all its parts carry compressive rather than tensile forces. If any member were placed in tension, as by wind stress or sinking of the foundation, collapse of the structure would follow.) The tensile strength of crustal rock is far from sufficient to pull a plate over the mantle below. If, alternatively, the plate were pushed, the compressive strength might be adequate but the plate would be expected to buckle (because the plate is so thin compared to its lateral dimensions). No evidence of buckling of the ocean floor plates has ever been detected.

The most reasonable explanation of the motion of the plates is that they are being carried along by the flow of the mantle rock immediately below. This mantle rock is at a high temperature and can flow relatively easily over long distances. The flow pattern is suggested in Figure 8.2: upwards under the ridges, outwards under the plates and downwards at the trenches. There must be a return flow deep in the mantle. This circulation pattern in the mantle suggests convection with the relatively hot material rising under the ridges and cooler material sinking under the trenches. Unfortunately, many difficulties stand in the way of a successful quantitative theory of mantle convection. The fact that the plastic properties of mantle rock vary with depth and are nonlinear (flow rate is not directly proportional to stress) means that the usual hydrodynamic equations of convection do not apply; there is the further complication of the phase changes which occur in upwards or downwards moving material in response to pressure and temperature gradients in the mantle. Although the nature of the motions within the mantle remains obscure, their reality cannot be doubted.

CONTINENTAL DRIFT

Because continental rock is on top of the plates, the relative positions of the continents must be changing. For example, North and South America must be moving away from Europe and Africa since these pairs of continents are on opposite sides of the mid-Atlantic ridge. This relative motion is known as *continental drift*. The evidence which originally suggested continental drift is, first, the remarkably close fit which may be made between some continental boundaries, and second, the coincidence of certain geologic structures when such a fit is made. Such observations are suggestive, but a convincing demonstration had to await the proof of sea floor spreading. The measurement of sea floor spreading rates now shows not only that continental drift is occurring, but that the size of continental separations so produced are of the magnitude actually observed. At a spreading rate of 5 cm/yr the present spacing between the Americas and Europe/Africa would require about one hundred million years to develop; this is well within the amount of time available, since the continental masses are at least 3000 million years old.

In testing for a fit between the continents it is not the shorelines which should be matched but, rather, the outline of the continental margins (which may be taken conveniently as the 100 meter curve on a hydrographic chart). The fit that can be made among the margins of the Americas, Europe and Africa is shown in Figure 8.13. There is a remarkable coincidence of the continental margins. Other continental

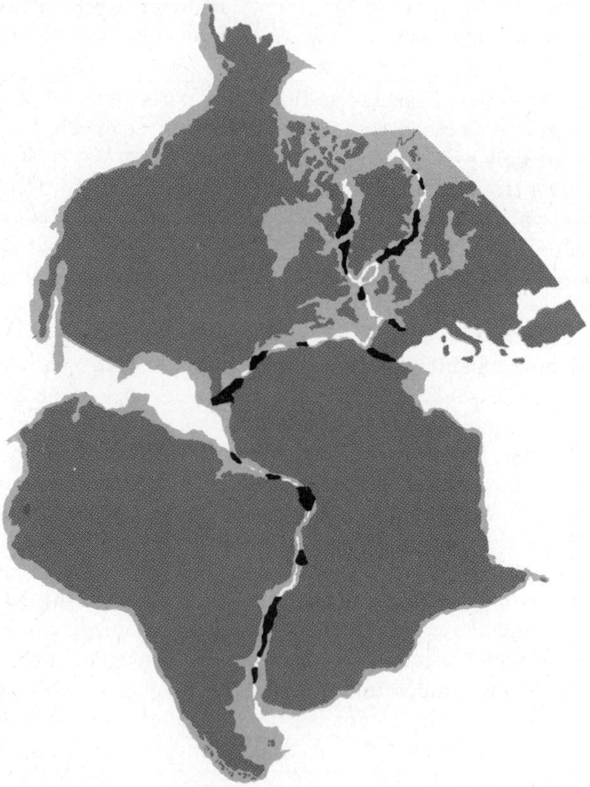

Figure 8.13 The best fit between the continental margins of the Americas, Europe, and Africa. Dark areas show overlap, white areas indicate gaps. After Bullard, Everett and Smith *Scientific American* (April, 1968).

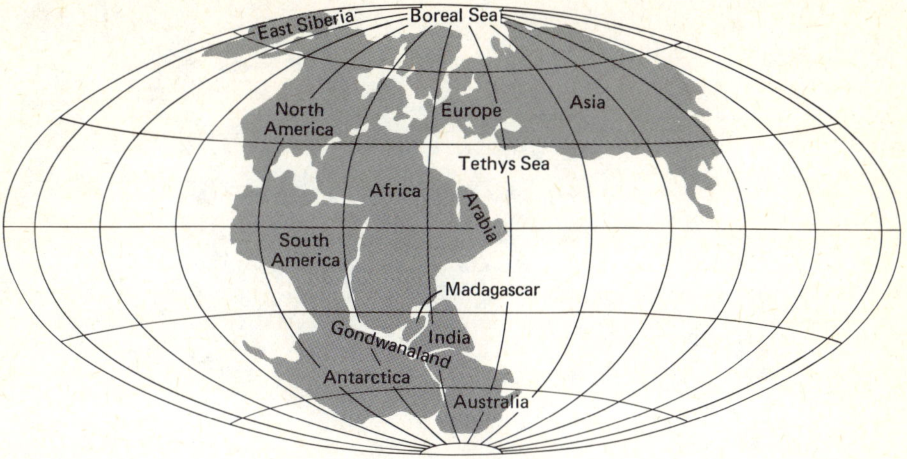

Figure 8.14 One possible reassembly of the continents. This is how a map of the world would have looked 180 million years ago.

areas can also be maneuvered about and fitted together like the pieces of a jigsaw puzzle in an attempt to find possible earlier continental configurations. One fit which can be made is shown in Figure 8.14. It indicates the existence at an earlier time of a single continental mass, the breakup of which was the beginning of the formation of the present continental array. There is evidence that breakup and reassembly of the continents has occurred a number of times in the course of the earth's history.

A number of tests can be applied to proposed continental reconstructions. There should be continuity of those parts of the continental structures which were formed before the date of breakup. There should be a continuity of evidence of climatic fluctuations (such as glacial ages) before the breakup date. And there should be found in the fossil record uniformity in the evolutionary development of the animal species on the land areas that were connected, followed, when breakup occurred, by evolutionary divergence. Positive evidence of all three types has been found. Such evidence is not by itself confirmation of continental drift, but it is difficult to explain in other ways. There is also evidence for continental drift from paleomagnetism. Suppose that the directions of residual magnetization in rocks formed at the same time but on different continents are compared. If the continental positions have remained fixed, all of these rocks should indicate the same position for the north and south magnetic poles. But observations show that the pole directions indicated by rocks on different continents do not coincide. Coincidence can be attained only by reorienting the continental blocks. Again, the simplest explanation is continental drift.

The pulling apart of a continental mass implied by the existence of continental drift must certainly have done violence to the continental structure in the zone of separation; it should be possible to find evidence of the fracturing which occurred. Evidence has indeed been found both of ancient fractures and of fracturing in progress today. The remains of tension cracks dating from the time (Triassic, see Appendix II) of the initiation of break-off of the Americas from Europe and Africa are found along the east coast of North America. Each crack is now a valley; they are shown in the map of Figure 8.15. The Connecticut Valley is typical. The cross section, Figure 8.16, shows that, under tension stresses in the crust, a fracture surface (called a *fault*)

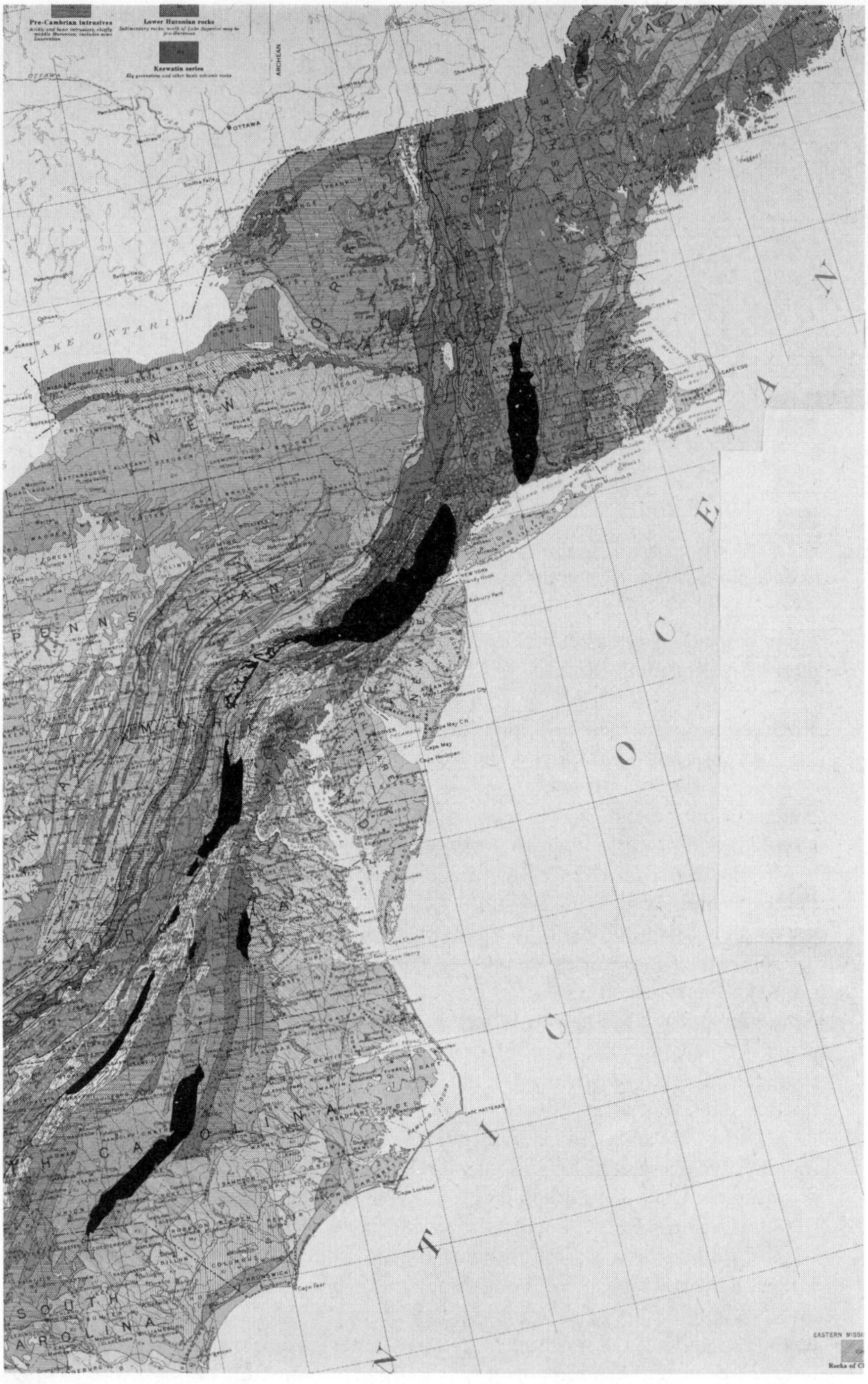

Figure 8.15 Location of the rift valleys on the East Coast of the United States formed at the time of the start of the breakup of the continental mass shown in Figure 8.14. (Based on the Geologic Map of the U.S., U.S. Geological Survey.)

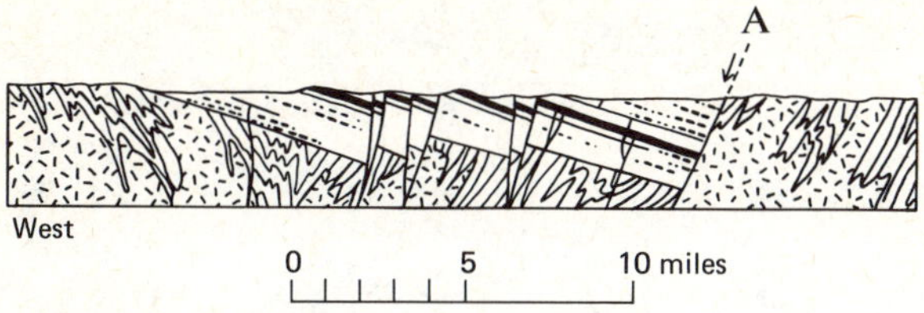

Figure 8.16 Cross section of the Connecticut rift valley. Lava flows are shown as dark layers. After King, *Evolution of North America*, Princeton U. Press (1969).

dipping downwards at an angle of about 70° formed. This is at A in the figure. The continental rock sank to the west of the fault forming a lowland bordered by high mountains to the east. Swift rivers running off these highlands carried sediments which were deposited layer upon layer in the new valley. The downwards inclination of these layers of sediment, which resulted from continuous motion on the fault, can be measured in the field today. Because of the overall tensile stress in the region, the resistance of the crust to penetration by hot lava from below was reduced. Basalt lava flowed upwards through tensile cracks to form the lava flows indicated in the figure. The present topography of the valley is the result of erosion of the relatively weak sediments leaving the lava flows exposed as steep-sided hills.

Active separation of continental areas is occurring today in the vicinity of the Red Sea and the Gulf of Aden (Fig. 8.17). A series of rift valleys like those of North America, but larger and much more recently formed, extend down the east coast of Africa. There is abundant evidence of recent volcanic activity and of pieces of mantle material carried to the surface. The occurrence of hot, highly saline water in pockets along the bottom of the Red Sea is evidence of the rapid upwards flow of heat from the mantle below.

Since continents are pulled apart by forces resulting from motions in the mantle, it is to be expected that they are also pushed together. The meeting place should be marked by structural features associated with compressive forces. The breakup of the continental pattern of Figure 8.14 suggests the northward movement of the Indian subcontinent into the Asiatic land mass. The resulting compressional features are the Himalaya Mountains. These examples suggest that plate motion is one of the forces which shape the complex structure of the continents, affecting the distribution of mineral resources and the character of the landscape. In the next section we look more carefully into continental structure.

CONTINENTAL ROCK

The ocean basins are geologically modern in that nearly all their structural features are the result of dynamic processes now active. (The Darwin rise in the southwest Pacific is thought to be the remains of a now inactive ocean ridge; such inactive structures are rare in the ocean basins.) Little on the sea floors is more than 80 million years old. On the continents some rock structures have remained unchanged for 3400 million years. In other places, the participation of continental rocks in repeated episodes of mountain building is revealed by the geologic record. Indeed, continental history is so rich in events that it is often difficult to separate those which are indicative of the general pattern of continental evolution. Much of the description of the geology of the continents was worked out long before the concepts of plate tectonics. The evolution of the continents was described in terms of a series of arbitrary upwards and downwards motions of land masses. We recognize today that these ver-

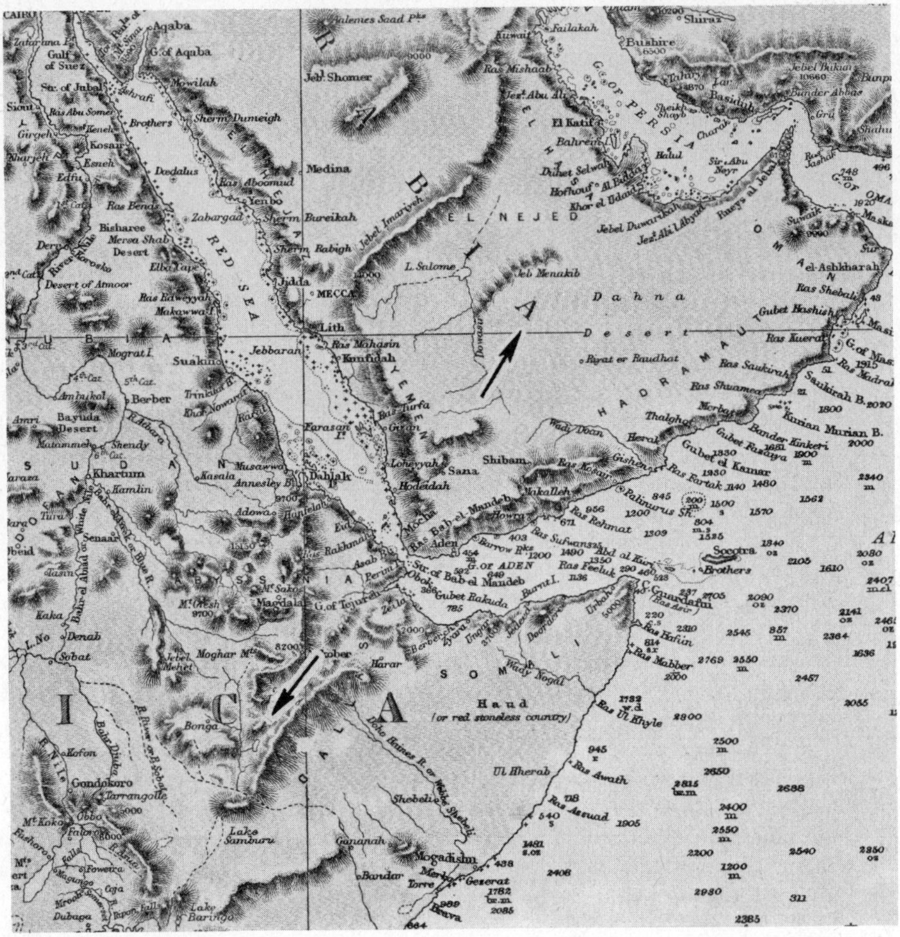

Figure 8.17 The Red Sea and Gulf of Aden are being enlarged by the separation (indicated by arrows) of Arabia from Africa. (Reproduced from BA chart number 2483 with the sanction of the Controller, H.M. Stationery Office and of the Hydrographer of the Navy.)

tical motions are consequences of the much larger horizontal movements of plates in the crust and upper mantle. The relationship between continental structure and plate tectonics is just starting to be worked out and will continue to occupy the attention of geologists and geophysicists for many years to come.

The rocks of which the continents are composed originate in one of three ways: by the deposition and consolidation of sediments (*sedimentary rock*), by the solidification of lava (*igneous rock*), or by the subsequent burial of sedimentary or igneous rock to successively greater depths where they are exposed to increasing temperature and pressure. With deep burial the structure of the original rock is altered to that of *metamorphic rock*. Depending on the depth of burial, various degrees, or grades, of metamorphism may be produced. The classification of rock as igneous, sedimentary or metamorphic is based on the fundamental processes of continental evolution that have been identified: igneous (volcanic) activity, formation of sediments by erosion and deposit, and burial of rocks formed at the surface under subsequent deposits of sediment.

Most sedimentary rocks are deposited from relatively still water and so are laid down as extensive, flat sheets. Changes in the conditions of sedimentation from time to time result in formation of layers of sediment of different composition or texture (Fig. 8.18). Further deposit of sediment compacts the material below and the conversion of loose sediment to rock begins. Obviously sedimentary in origin are sandstones (sand grains cemented together by iron oxide) and mudstones (solidified mud) formed by deposit from slow-moving water. Other sedimentary rocks are deposited from solutions. *Evaporite* beds are composed of sea salts left behind after the evaporation of salt water. Much more abundant are the carbonate rocks, such as limestone, composed of the remains of marine animal shells (which are mostly made of calcium carbonate, $CaCO_3$). The chalk which makes up the white cliffs of Dover is simply compacted shells of tiny marine organisms. As might be expected, this rock is soft and easily cut back by the sea.

Characteristic changes accompany metamorphism of sedimentary rocks. Exposed to high temperature and pressure, limestone recrystallizes and becomes *marble;* sandstone recrystallized becomes *quartzite* (Fig. 8.19). Mudstone becomes *shale* or, if metamorphosed under a compressive stress which aligns all of the constituent clay minerals in one direction, *slate* (Fig. 8.20). This metamorphism is of low enough degree that, although the properties of the sedimentary rock are changed, its origin is not obscured. Thus, the bedding of the original layers of mud can still be seen in the slate of Figure 8.20. When the metamorphic grade is high enough to form *gneiss* the origin of the starting rock may no longer be evident. At the highest grade of metamorphism partial melting of the rock begins, the low melting point constituents begin to

Figure 8.18 Exposure of sedimentary rock near Berlin, Connecticut, showing the sequence of layers which results from changes in the conditions of deposit of sediments.

Figure 8.19 Photomicrograph of the rock quartzite showing recrystallized quartz grains.

Figure 8.20 A block of slate from the Martinsberg formation in Northern New Jersey. Flat faces are formed by cleavage parallel to the flat faces of the constituent clay mineral grains. The original layers of sediment are marked by the light-colored bands. Metamorphism of the sediment under stress has caused a reorientation of the clay minerals originally parallel to the color bands to that now shown by the cleavage. Photograph by A. Coleman.

make their way out of the rock and little is left of the original structure or composition, as in Figure 8.21. Metamorphism of igneous rock produces similar changes in structure and texture though these rocks tend to be less affected than the sedimentary rocks by the lower degrees of heat and pressure.

Often indicative of the origin of sedimentary rock is the occurrence in it of fossil remains of plants and animals. Relative ages of strata can often be established by the occurrence of fossil life types which can be placed in the evolutionary sequence. High grade metamorphism destroys the fossil evidence, unfortunately, so this important guide to the origin of rock is not available where the rock has been greatly altered by exposure to heat and pressure.

CONTINENTAL STRUCTURES

The continents are composed of igneous, sedimentary and metamorphic rock organized in structures which range from simple, flat beds of sediment to complex mountains. Large parts of central North America are covered by flat-lying sedimentary deposits arranged in the order of their deposition and essentially undisturbed since that time. Small, vertical movement of the underlying rock (the *basement*) has in some places produced domes (upwards motion) and basins (downwards motion). These structures are simple; the Michigan basin is an example. A geologic map, which displays the sequence of formation times of rock seen at the surface, is shown for this area in Figure 8.22. The map shows that the youngest sediments are found in the center of Michigan and these younger rocks are surrounded by successively older strata. The cross section of the structure shown in this geologic map is drawn in Fig-

Figure 8.21 Rock partially melted during metamorphism at very high temperature. Photographed at East Lyme, Connecticut.

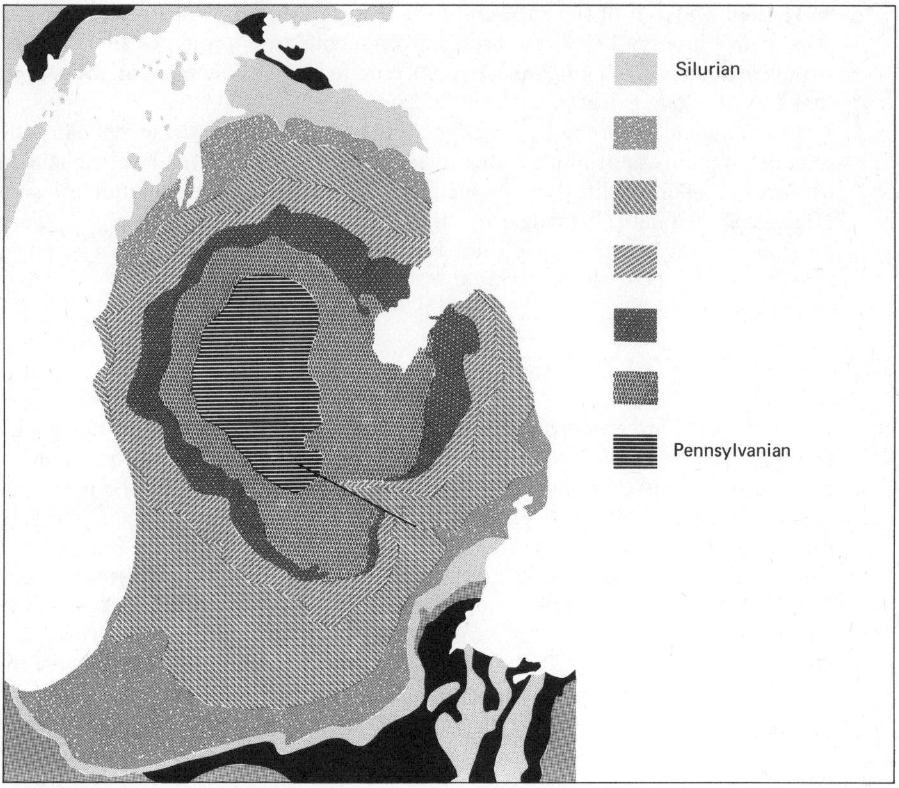

Figure 8.22 Geologic map of Michigan showing the ages of rocks exposed at the surface. Relative ages are shown by the scale from Silurian (oldest) to Pennsylvanian (youngest). Based on the Geologic Map of the U.S., U.S. Geological Society.

ure 8.23. What is now Michigan was once the floor of a shallow sea. As sediment layers were deposited, the sea floor sank in the middle of what is now the Michigan peninsula. Compaction and the passage of time converted the sediments to sedimentary rock; erosion has cut through the sedimentary pile to expose the present surface with its characteristic sequence of strata.

A range of continental mountains is a great deal more complex in structure, but a regular progression of structures can be recognized as common to many mountain

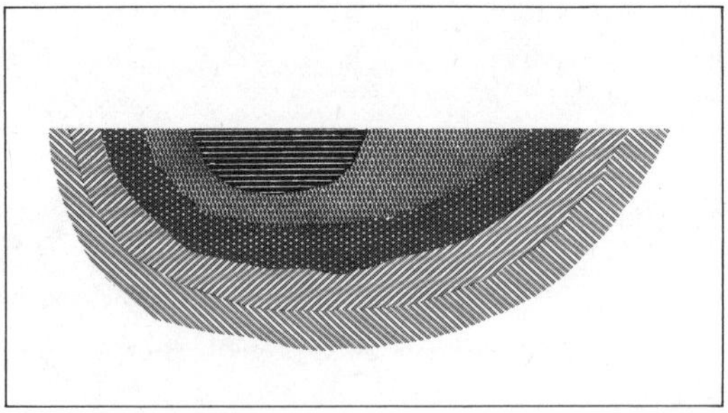

Figure 8.23 Schematic cross section of the area shown in Figure 8.22.

belts. The Appalachians are a good example of this progression. To the northwest of the Appalachians are rocks which have remained undisturbed for more than 500 million years. These occupy an area known as the *Canadian shield;* it covers much of southeastern Canada. The ancient rocks of the shield dip away to the southeast where they are covered with increasingly thick layers of more recently deposited sedimentary rock. A cross-sectional view of these layers of sedimentary rock along a line running southeastwards across Pennsylvania is shown in Figure 8.24. The structure of flat sedimentary layers in the section through Pennsylvania ends abruptly just west of the town of Tyrone in the imposing escarpment known as the Allegheny Front. Southeastwards from the front into the mountains there is increasing deformation, faulting, and metamorphism of the sedimentary strata. The sequence of structures can be described in terms of four provinces, each characterized by a definite type of structure.

The Foreland Province consists of nearly flat beds of sedimentary rock, each bed being younger than the one immediately below. The oldest were deposited about 570 million years ago (Cambrain times, see Table 8.1 and Appendix II) and the youngest, about 280 million years ago. These layers of sedimentary rock are in many places rich in coal and oil formed from the abundant plant and animal life that flourished when the sediments were deposited.

The *Valley and Ridge Province* begins at the Allegheny Front, where the sedimentary layers of the Foreland are deformed by horizontal compression. The shortening of the layers of sedimentary rock is accomplished in large part by folding, a crumpling up of the layers similar to the folding of a rug pushed on one side. This folding is clearly shown in Figure 8.24. Where folded strata reach the surface they are exposed to erosion; the more resistant rock is left standing higher than the more easily eroded layers. The resultant topography is a series of parallel ridges composed of resistant rock separated by valleys floored with soil derived from the less resistant rock of the softer strata. The cross section of the ridges is often revealed in the wind and water gaps which have been cut through the mountains by rivers (Fig. 5.3). (In a "water gap" the river still flows, in a "wind gap" there is no longer a stream present.) Figure 8.25 shows the cross section of Kittatiny Mountain in the Delaware Water Gap. Erosion has cut away the top of a large fold leaving the sedimentary layers now visible in the gap inclined upwards at about 45°. The upturned edges of these strata form a long, straight mountain.

The occurrence of folding implies that the sedimentary strata were, at the time of folding, deeply enough buried to have plasticity sufficient to allow gentle bending of the rock layers without breaking. The relative displacements which must occur between layers during bending are accommodated by sliding at the layer boundaries, with the result that in the folds the thickness of the strata does not change. This is clearly seen in the folds at Stone Mountain and Tuscarora Mountain on the Pennsylvania section of Figure 8.24. When the plasticity of the sedimentary rock is insufficient to allow the shortening to be accomplished by folding, *thrust faults* occur. A thrust fault is illustrated in Figure 8.26. A crack inclined at about 45° to the surface has formed in the relatively brittle rock. Under horizontal compressive forces the layers of sedimentary rock have been pushed over one another, sliding on the crack faces. A few thrust faults can be seen in Figure 8.24, for example, between Stone Mountain and Jacks Mountain, but these faults are more common in the southern Appalachians. The displacement of the strata on a thrust fault may be as much as 10 kilometers; the fault may be as much as 200 kilometers long.

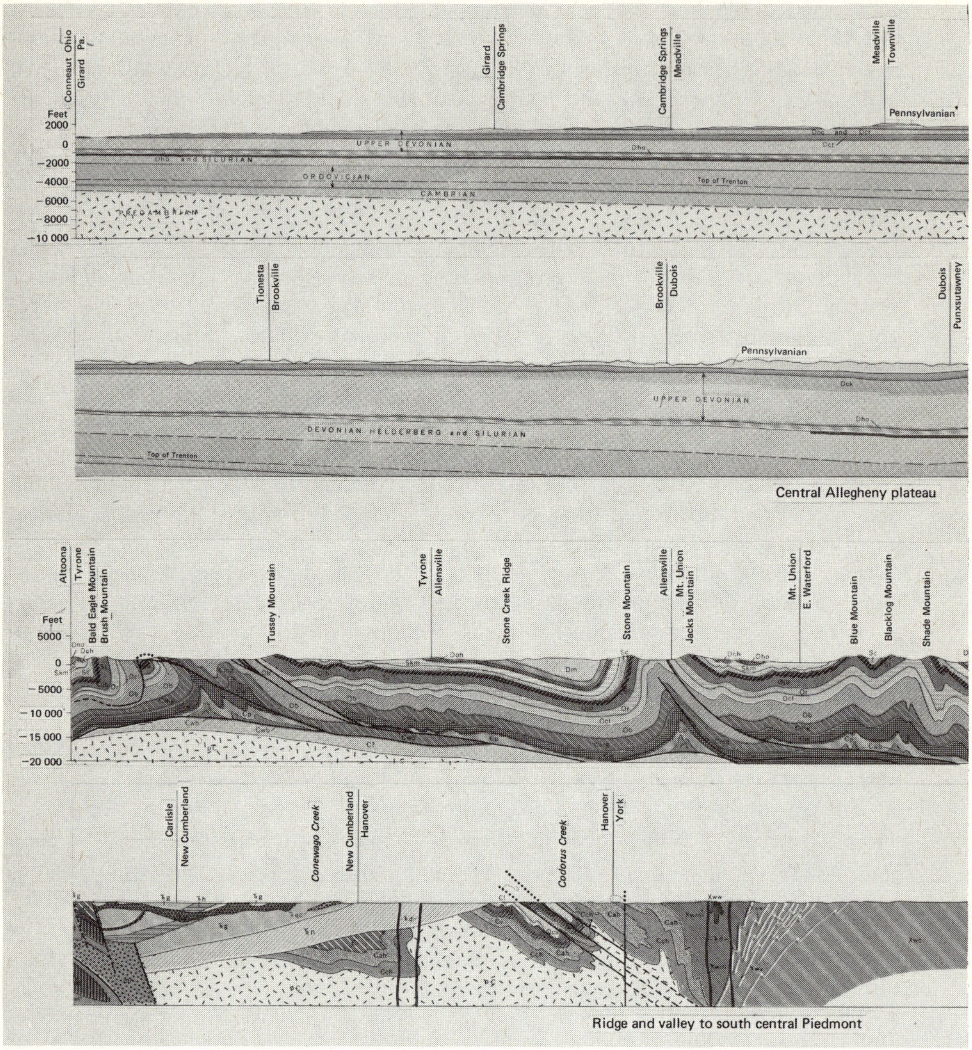

Figure 8.24 Part of the geologic map of Pennsylvania showing a cross section through the Appalachian mountain provinces. (From Geologic Map of Pennsylvania, Pennsylvania Geological Survey.)

In the *Blue Ridge Province* there is an increase in the intensity of the compressive deformation compared to the Valley and Ridge. The fold amplitude in the Blue Ridge becomes large enough to bring to the surface the deepest, and therefore oldest, strata. The character of the folding is changed and is now that of a material which flows very easily. There is a relatively great thickening of strata along the axis of the folds. This can be seen easily in the folds at South Mountain in Figure 8.24. To the south, the Blue Ridge Province of the Appalachians includes the Great Smoky Mountains, in the north, the Green Mountains of Vermont.

The change in the character of the folds in the Blue River Province foreshadows

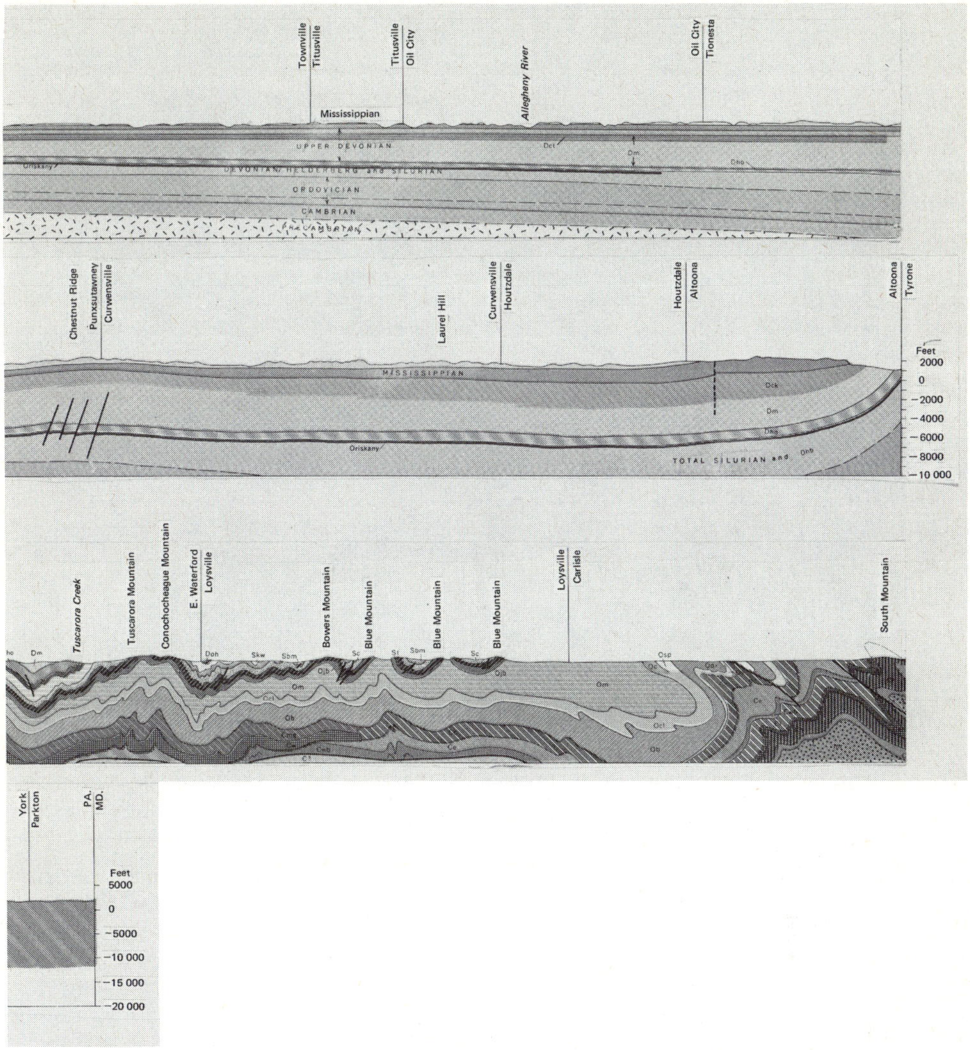

the intense metamorphism found on further penetration into the mountain chain. (Note that there is not necessarily a correlation between the intensity of deformation and mountain topography. In parts of the Appalachians the Valley and Ridge may show more rugged topography than the most intensely deformed land further east.) In the *Crystalline Province* folding becomes intense and complex, simple linear features are not so apparent, and the deformation has been accomplished at such high temperature and pressure that the rock loses its sedimentary character. Partial fusion of the sediments has occurred in the intense metamorphic zone with the fluids so released making their way to the surface, sometimes through channels like those shown in Figure 8.27, and sometimes as much larger masses of fluid which upon solidification become *granite* or *pegmatite*.

In the south the Crystalline Province of the Appalachians has been removed by

Figure 8.25 Section of Kittatiny Mountain seen at the Delaware Water Gap.

erosion; in the north it makes up much of central New England, notably the White Mountains of New Hampshire. Further to the east the intensity of the mountain-forming forces declines; the rocks are soon lost to view under the Atlantic coast so that structural details are not so well known, but deformed and metamorphised rock extends well out on the continental margin.

The parallel provinces of the Appalachians run along the eastern seaboard of North America, passing to the north under the Gulf of St. Lawrence and then across Newfoundland. The Appalachians were formed well before the separation of North America from Europe and Africa (which occurred after the initial fracturing evidenced by the Triassic valleys). The northern end of the Appalachians should therefore be found today on the other side of the Atlantic. In fact, most of the elements of the Appalachian structure can be recognized in the Caledonian and Hercynian mountain belts which strike across England and Scotland. Exact matching is not possible because a 300 km long segment of the mountain chain is now buried

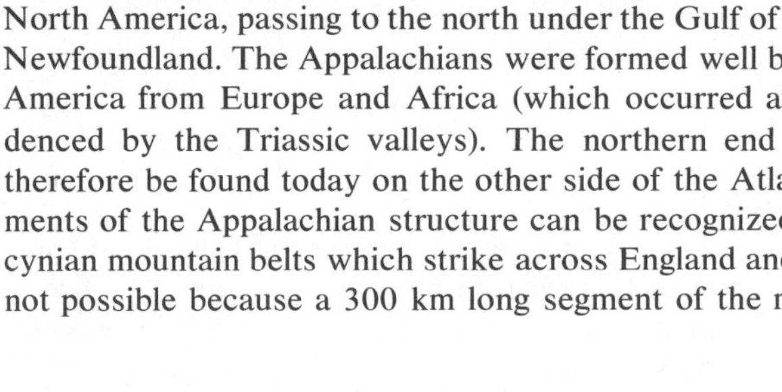

Figure 8.26 Formation of a thrust fault in strata compressed horizontally.

Figure 8.27 Veins of pegmatite (solidified melt) now mark channels by which melt formed in high temperature metamorphism was carried upwards. Photographed near Essex, Connecticut.

under the sediments of the continental shelf, but there is little doubt of the continuity of the structures.

The general features of the Appalachian structure described above are found in many other mountain systems—in the Rockies, for example—where the sequence of provinces progresses westwards. The similarity of structure suggests similarity of mountain-making forces around the world. The Appalachians are unusual in having a fairly uniform lateral development over their great length. Worn down by erosion, their internal structure is revealed in section. In their youth these mountains must have looked like the Rocky Mountains or the Alps.

MOUNTAIN BUILDING

The large scale of the deformation required to form a mountain range suggests the origin of mountain building forces in plate motion. Another suggestive fact is that the young mountain belts with seismic activity are found today at the margins of plates which are in collision. Thus the Andes are at the collision boundary of the East Pacific and American plates, the Himalayas, at the juncture of the Indian and Asian plates. Examination of the processes which have formed mountain ranges shows two general cases. In the first case, oceanic crust of one of the colliding plates is thrust under a continental margin on the other plate, as in the Andes. In the second case,

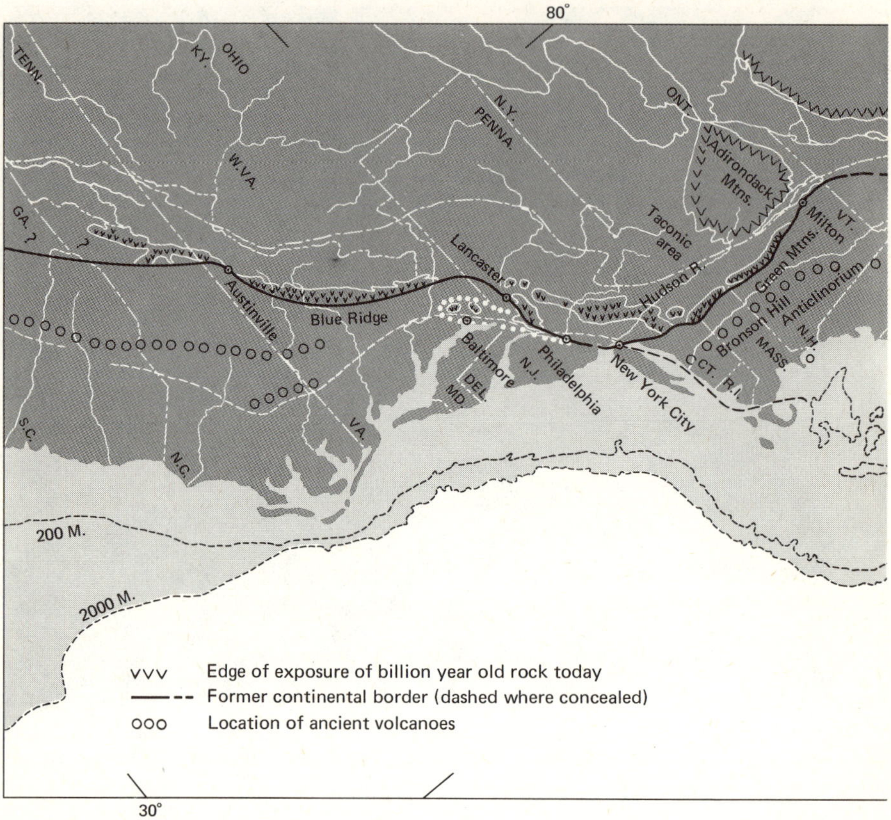

Figure 8.28 Map showing the continental border of North America in Cambrian times and the zones of greatest volcanism. After Rodgers.

continental margins on two plates are driven together, as in the Himalayas. The first case is more common and was evidently responsible for the formation of the Appalachians.

In the case of the Appalachians the centinental margin existed over 570 million years ago (before Cambrian times) and was composed of the rock described above as the "basement." (This rock had participated in earlier episodes of mountain building and had been worn down by the beginning of the Cambrian times to a nearly smooth surface.) The presumed location of this continental margin is shown in the map of Figure 8.28. Collision with a plate carrying oceanic crust turned the advancing, heavier oceanic crust down under the continental edge. This downturning of oceanic crust formed a trench in the sea floor and carried marine sediments under the continental crust. Decomposition of the sediments thrust deep into the earth at the trench started the release of lava which, making its way to the surface, began to form an arc of volcanic islands. This is the beginning stage of mountain formation, shown in Figure 8.29. The lava released from these volcanoes produces silica-rich rock of the type which makes continental crust. (Volcanism of the ocean ridges produces silica-poor rock, basalt, characteristic of material brought up from the upper mantle.) Thus, mountain formation resulting from plate collision caused continental growth. After

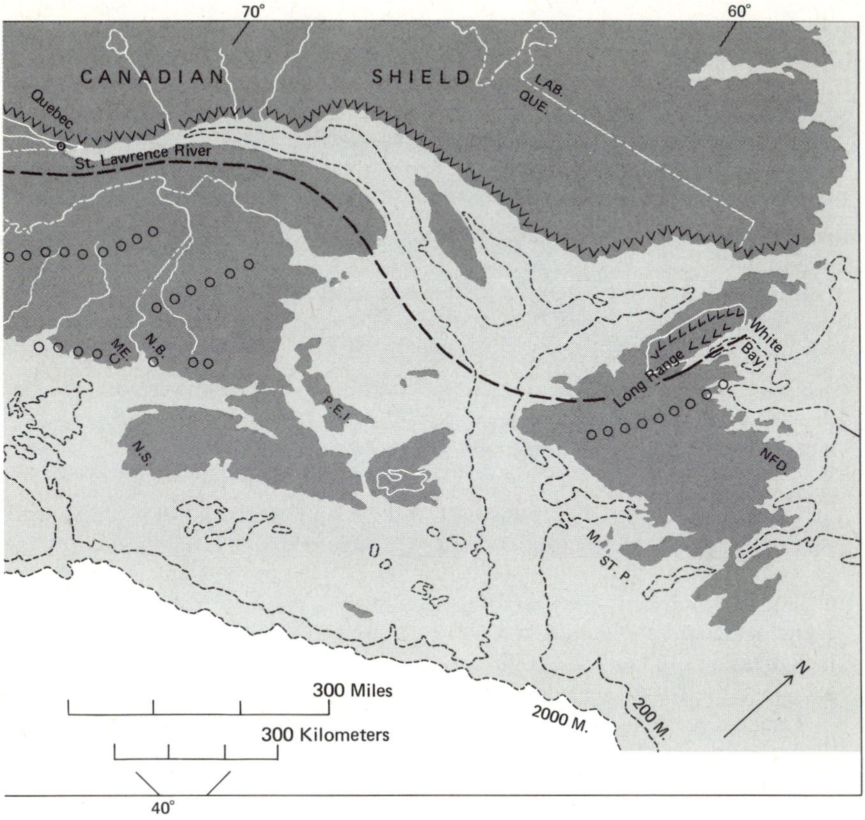

formation of the volcanic islands marine sediments began to accumulate in the relatively shallow water between the island arc and the old continental margin. The earliest of these sediments (Cambrian age) were derived from the continent to the west, but later (in the Ordovician age), coarse sediments began to enter this shallow basin from the east. Discovery of such sediments today is evidence of the formation in the past of new, high land to the east—the island arc. The structure at this stage was a shallow sea bounded on the ocean side by emergent islands. Such a structure is found today in the Sea of Okhotsk. Metamorphism and folding were most intense near the volcanoes of the island arc and between the arc and the trench to the east. This zone is now recognized as the Crystalline Province.

As the volcanism continued, the island arc grew up into high land and the flow of sediment continued westwards, eventually reaching the present edge of exposed basement rock shown in Figure 8.28. The compressive thrust from the opposing motion of the two plates, and from the injection of much molten lava into crust, produced buckling (folding) and faulting of the sedimentary layers to the west as the island arc structure was compressed. The intensity of these compressional features declines to the west because the thin layers of sedimentary rock are unable to transmit compressive forces over large distances; the force is rapidly dissipated through friction to the basement below. Deformation of the sediments stopped at what is now the Appalachian Front.

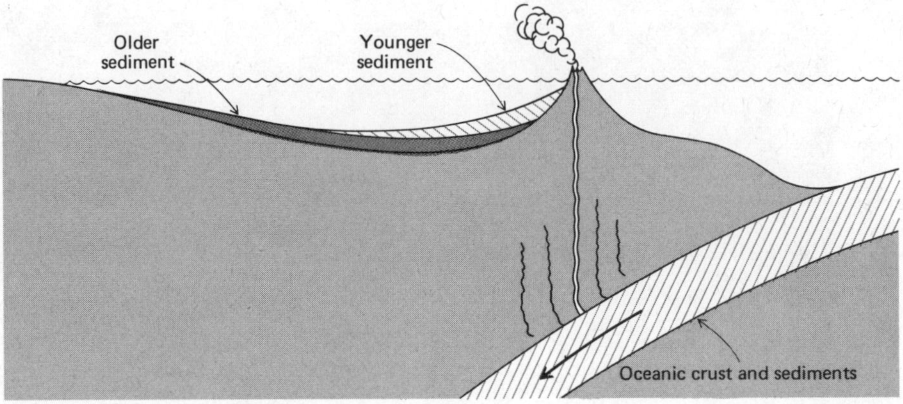

Figure 8.29 Early formative stage of the Appalachians, showing the structure developed where a plate bearing a continent collides with a plate of oceanic crust.

Events which occurred to the east of the island arc cannot now be studied in the Appalachians because these rocks are now covered by thick layers of recently formed sediments and by the sea. They can be seen in the Franciscan Mountains of California: Sediment has been carried down rapidly to great depths by the descending oceanic crust and was there exposed to a characteristic metamorphism of high pressure but low temperature. Faulted structures of great complexity have been produced by the stresses developed by the downturning layer of crust.

Eventually the mountain-building activity resulting from plate collision begins to decline. Whether this results from a shift in the position of the plate, carrying the continent away from the trench, or from a sudden shift in the underlying pattern of mantle motions is not clear. While evidence of two major periods of mountain-building activity in the Appalachians separated by a period of about 50 million years has been found, no one has yet discovered what pattern of plate motions caused these

Table 8.1 Summary of Principal Events in Appalachian History

Time (millions of years ago)	*Period*	*Events*
1.5	Quaternary	Glaciation and subsequent development of coastlines.
65–1.5	Tertiary	Intensified erosion generating much of the present topography of the mountains. Continued deposition of sediments to the east.
136–65	Cretaceous	Erosion, deposition of sediments to eastward.
136–193	Jurassic	Continued erosion.
193–225	Triassic	Formation of rift valleys, continued erosion of the high standing mountains.
280–225	Permian	End of tectonic activity.
310–280	Pennsylvanian	Erosion and sedimentation active.
310–280	Mississippian	Erosion and sedimentation active.
395–345	Devonian	Second period of mountain building (Acadian) followed by erosion.
435–395	Silurian	End of mountain building; erosion active.
500–435	Ordovician	First episode of mountain building (Taconian).
570–500	Cambrian	Deposition of sediment from westward.

separate events. In fact, little is yet known of plate patterns which existed before the last continental break-up (200 million years ago).

After the end of the mountain-building period, events in the history of the Appalachian region include its exposure to tensile forces beginning with the Triassic (previously discussed in connection with the rift valleys), continued erosion of the high mountains until a nearly plane surface was formed, and some uplift of the region followed by erosion which has cut the present topography. A simplified record of the sequence of events is summarized in Table 8.1.

The theory of plate tectonics provides a means of understanding the sequence of events which forms a chain of continental mountains. It even offers the possibility of predicting future events to some extent. Mountain chains in various stages of development can be found around the world. Comparison of the older with the newer suggests what the future of the newer ones will be. But there are a great many detailed geologic observations still to be drawn into the general theory. And there is yet the central, unsolved problem of what controls the plate motions. When this is understood the theory of mountain building and of continental structures will be much nearer completion. In the meantime, practical applications of the developing theory can be made in searching for ore minerals, or in better understanding the country one passes through on a transcontinental trip.

Further Readings

CLARK, S. P., *The Interior of the Earth.* Englewood Cliffs, N.J.: Prentice-Hall, 1971.

FRASER, R., *The Habitable Earth.* New York: Basic Books, 1964.

GASKELL, T. F., *Physics of the Earth.* New York: Funk and Wagnalls, 1970.

KING, P. B., *The Evolution of North America.* Princeton, N.J.: Princeton University Press, 1959.

SHELTON, J. S., *Geology Illustrated.* San Francisco: Freeman, 1966.

STRANGEWAY, D. W. *History of the Earth's Magnetic Field.* New York: McGraw-Hill, 1970.

TARLING, D. H., and M. P. TARLING, *Continental Drift: A Study of the Earth's Moving Surface.* London: G. Bell, 1971.

Appendix I Definitions and Units

DEFINITIONS

The fundamental physical quantities are mass, length, and time. The concepts of length and of time are familiar from everyday experience; mass is a measure of the resistance that an object initially at rest offers to being set in motion. In the metric system of units, length may be measured in centimeters, time in seconds, and mass in grams. All other physical quantities can be defined in terms of mass, length and time and, consequently, measured in grams, centimeters, and seconds. Thus:

Velocity, v, is the distance, d, covered in unit time, t,

$$v = d/t = \text{cm/sec}$$

To completely describe the velocity of an object, the direction in which it is moving as well as its speed must be specified. Thus, velocity is a *vector* quantity; that is, a quantity characterized by both a magnitude (in this case the speed) and a direction. Velocity may be represented by an arrow pointing in the direction of motion, with a length proportional to the speed.

Acceleration, a, is the rate of change of velocity. The acceleration required to bring an object from rest to velocity v in time t is

$$a = v/t = (\text{cm/sec})/\text{sec} = \text{cm/sec}^2$$

Like velocity, acceleration is a vector quantity.

Force, F, must be applied to an object of mass m to make it accelerate. Its magnitude is

$$F = ma = \text{gm cm/sec}^2 = \text{dynes}$$

A force of one dyne will accelerate a mass of one gram at one centimeter per sec.^2 Force is a vector and its direction is the same as that of the resultant acceleration.

Because velocity and acceleration are vectors, force must be applied to change the direction in which an object moves. The most familiar example is *centrifugal force.*

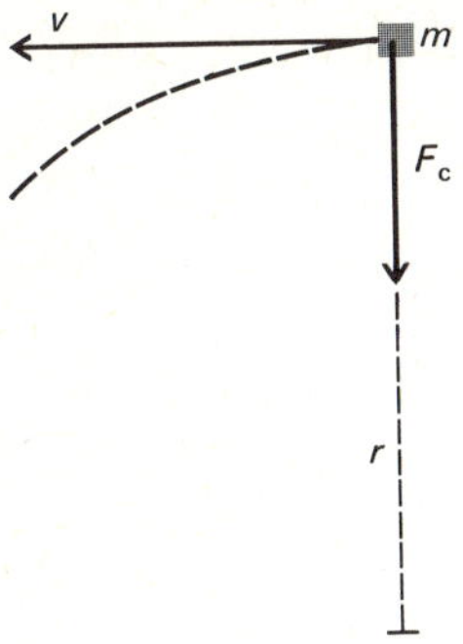

Figure I.1 The mass m is moving to the left with velocity v. In order to make it follow the circular path having radius r the centripetal force F_c must be applied.

To make a mass m follow a circular path of radius r, the centripetal force

$$F_c = mv^2/r$$

must be applied in a direction perpendicular to v (Fig. I.1) to balance the centrifugal force.

A fluid, such as the air in a pneumatic tire, exerts force in all directions on its surroundings. The *pressure, P,* of the fluid is the force per unit area:

$$P = F/A = \text{dynes/cm}^2$$

The pressure produced by a fluid, which acts equally in all directions, is called *hydrostatic* pressure. Nonhydrostatic pressure, as when two solid objects press against each other, is often called *stress*. It is also measured in units of dynes/cm^2.

Density, ρ, is the mass concentration, the amount of mass contained in a unit volume. Thus,

$$\rho = m/V = \text{gm/cm}^3$$

Note that the density of water is very nearly one gram per cubic centimeter.

Momentum is defined as the product of mass times velocity:

$$\text{momentum} = mv = \text{gm cm/sec}$$

The greater the momentum of a moving object, the more difficult it is to bring it to rest.

Work is done when force, F, is applied to move an object over some distance, d. Work is the expenditure of *energy, E,* which is defined as

$$E = Fd = \text{dyne cm} = \text{erg}$$

A useful physical rule is the law of *conservation of energy:* Energy can be transferred from one place or form to another, but its total amount must remain constant. For example, work is done in lifting a weight to some height above the ground. This increases the *potential energy* of the weight by the amount of work done. In falling to the ground an equal amount of energy is released and can be used to do work on something else.

Work can also be done in increasing the velocity of an object; the object then acquires *kinetic energy*. The kinetic energy (KE) of an object of mass m moving at velocity v is

$$KE = \tfrac{1}{2}mv^2$$

The kinetic energy is released when the moving object is stopped. Work done against frictional forces, as in rubbing sandpaper over wood, appears as *thermal energy,* manifested by a rise in the temperature of both the wood and the sandpaper. *Electrical energy* is created by the work required to move a metallic wire through a magnetic field. The law of conservation of energy applies to all interchanges among these different types of energy.

UNITS

The choice of grams, centimeters, and seconds as units is arbitrary. Other choices are possible, for example, length in feet or time in minutes. However, there is a growing tendency towards the universal use of the units of the metric system (grams, meters, seconds). In a few years' time it will be only in North America that metric units are not in common use. The significant advantage of the metric units is that they are a decimal system. For example, 1 meter = 100 centimeters, and 1000 gm = 1 kilogram. The prefixes used for multiples of metric units are listed in Table I.1.

Table I.1 Prefixes Indicating Multiples of Ten

1/10	deci-	10	deka-
1/100	centi-	100	hecto-
1/1000	milli-	1000	kilo-
10^{-6}	micro-	10^{6}	mega-
10^{-9}	nano-	10^{9}	giga-

Sometimes in numerical work it is convenient to use powers of ten instead of the prefixes. Thus

$$1{,}000{,}000 \text{ meters} = 10^{6} \text{ meters} = 10^{3} \text{ kilometers}$$
$$2{,}300{,}000 \text{ meters} = 2.3 \times 10^{6} \text{ meters} = 2.3 \times 10^{3} \text{ kilometers}$$

In geophysics numbers occur which are too large to be expressed in terms of the prefixes. For example:

$$\text{Mass of the earth} = 5.976 \times 10^{27} \text{ grams}$$

The relationship between metric units and common English units is given in Table I.2.

The metric system is not completely consistent in the use of prefixes: Some multiples of basic metric units have been given special names. Examples are:

$$10^{6} \text{ dynes/cm}^{2} = 1 \text{ bar}$$
$$10^{5} \text{ dynes} = 1 \text{ newton}$$
$$10^{7} \text{ ergs} = 1 \text{ joule}$$

One important exception to the universal use of metric units is made in this book. In many places the *nautical mile* is used in preference to kilometers. This is because the nautical mile (6076 ft, 1.852 km, slightly greater than a statute mile of 5280 ft) is equal to the length of one minute of arc of latitude. Thus distance in nautical miles can be read directly from the scale of latitude which is marked on maps and charts. The unit of speed based on the nautical mile is the *knot.*

$$1 \text{ knot} = 1 \text{ nautical mile per hour}$$

The knot is a particularly useful unit in working out time and distance problems on charts and maps.

Table I.2 English/Metric Equivalents

Length	
1 meter = 39.37 inch = 3.280 ft	
1 kilometer = 3280 ft. = 0.621 statute mile = 0.540 nautical mile	
1 inch = 2.54 cm	1 fathom = 6 feet = 1.83 meters
1 foot = 0.305 meter	1 statute mile = 1.61 km
1 yard = 0.914 meter	1 nautical mile = 1.85 km
Mass	
1 gram = 0.032 ounce = 0.00220 pound	
1 kilogram = 2.20 pounds = 32.2 ounces	
1 pound = 0.454 kg	1 ton = 907 kg
1 ounce = 28.3 gm	
Speed	
1 meter per second = 3.6 km/hr = 2.23 miles/hr = 1.94 knots	
1 mile/hr = 1.61 km/hr	1 knot = 1.85 km/hr
Pressure or Stress	
1 bar = 14.5 lb/in^2 = 0.987 atmosphere	
Energy	
1 joule = 10^7 ergs = 0.240 gram-calories	

Wind speed in knots is often expressed in terms of the Beaufort scale shown in Table I.3. The practical utility of this scale is that most observers with a little practice can estimate wind speed to the nearest Beaufort number without the aid of instruments.

COMBINATION OF VECTORS

A practical problem which often arises is to find the force that results from the application of two or more separate forces on an object. If the forces are not in the same direction, their resultant must be found by *vector addition.* This can be done easily by the construction illustrated in Figure I.2. Forces F_1 and F_2 act at point A. The sum of F_1 and F_2 is the diagonal drawn in the parallelogram having F_1 and F_2 as

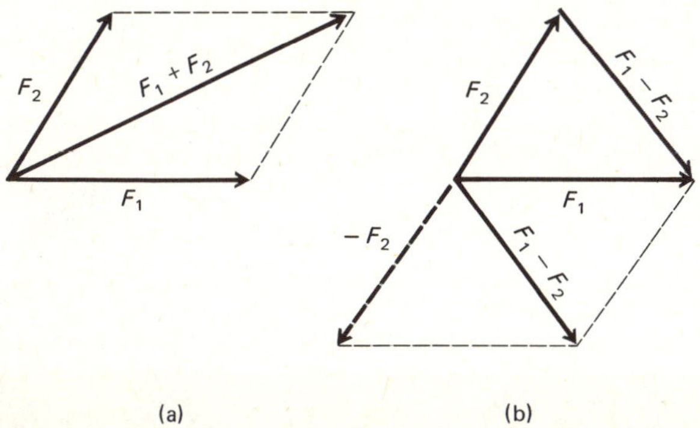

Figure I.2 (a) Addition of vectors F_1 and F_2 is accomplished by the construction shown. (b) Subtraction of F_2 from F_1 is achieved by adding $-F_2$ to F_1. An alternate, but equivalent, construction for finding $F_1 - F_2$ is shown.

Table I.3 Beaufort Scale with Corresponding Sea State Codes[a]

Beaufort number	Wind speed: knots	Wind speed: mph	Wind speed: meters per second	km per hour	Seaman's term	U.S. Weather Bureau term	Estimating wind speed: Effects observed at sea	Estimating wind speed: Effects observed on land
0	under 1	under 1	0.0–0.2	under 1	Calm		Sea like mirror.	Calm; smoke rises vertically.
1	1–3	1–3	0.3–1.5	1–5	Light air	Light	Ripples with appearance of scales; no foam crests.	Smoke drift indicates wind direction; vanes do not move.
2	4–6	4–7	1.6–3.3	6–11	Light breeze		Small wavelets; crests of glassy appearance, not breaking.	Wind felt on face; leaves rustle; vanes begin to move.
3	7–10	8–12	3.4–5.4	12–19	Gentle breeze	Gentle	Large wavelets; crests begin to break; scattered whitecaps.	Leaves, small twigs in constant motion; light flags extended.
4	11–16	13–18	5.5–7.9	20–28	Moderate breeze	Moderate	Small waves, becoming longer; numerous whitecaps.	Dust, leaves, and loose paper raised up; small branches move.
5	17–21	19–24	8.0–10.7	29–38	Freeze breeze	Fresh	Moderate waves, taking longer form; many whitecaps; some spray.	Small trees in leaf begin to sway.
6	22–27	25–31	10.8–13.8	39–49	Strong breeze		Larger waves forming; whitecaps everywhere; more spray.	Larger branches of trees in motion; whistling heard in wires.
7	28–33	32–38	13.9–17.1	50–61	Moderate gale	Strong	Sea heaps up; white foam from breaking waves begins to be blown in streaks.	Whole trees in motion; resistance felt in walking against wind.
8	34–40	39–46	17.2–20.7	62–74	Fresh gale	Gale	Moderately high waves of greater length; edges of crests begin to break into spindrift; foam is blown in well-marked streaks.	Twigs and small branches broken off trees; progress generally impeded.
9	41–47	47–54	20.8–24.4	75–88	Strong gale		High waves; sea begins to roll; dense streaks of foam; spray may reduce visibility.	Slight structural damage occurs; slate blown from roofs.
10	48–55	55–63	24.5–28.4	89–102	Whole gale	Whole gale	Very high waves with overhanging crests; sea takes white appearance as foam is blown in very dense streaks; rolling is heavy and visibility reduced.	Seldom experienced on land; trees broken or uprooted; considerable structural damage occurs.
11	56–63	64–72	28.5–32.6	103–117	Storm		Exceptionally high waves; sea covered with white foam patches; visibility still more reduced.	
12	64–71	73–82	32.7–36.9	118–133				Very rarely experienced on land; usually accompanied by widespread damage.
13	72–80	83–92	37.0–41.4	134–149				
14	81–89	93–103	41.5–46.1	150–166			Air filled with foam; sea completely white with driving spray; visibility greatly reduced.	
15	90–99	104–114	46.2–50.9	167–183	Hurricane	Hurricane		
16	100–108	115–125	51.0–56.0	184–201				
17	109–118	126–136	56.1–61.2	202–220				

[a] Note: Since January 1, 1955, weather map symbols have been based upon wind speed in knots, at five-knot intervals, rather than upon Beaufort number. From Bowditch, *American Practical Navigator*, through courtesy U.S. Naval Oceanographic Office.

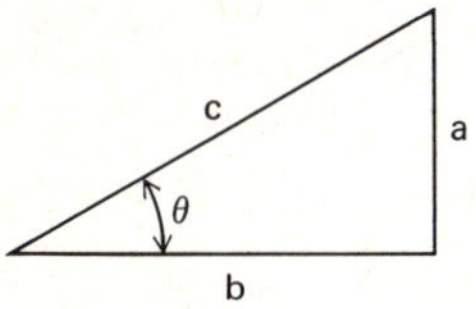

Figure I.3 Triangle used to define the trigonometric functions of the angle θ.

adjacent sides. The difference $F_1 - F_2$ is found by adding $-F_2$ to F_1. This construction is also shown in the figure.

In calculations involving vectors, it is often helpful to use trigonometric functions. These are simply the ratios of the lengths of the sides of right triangles. The definitions of the sine, cosine, and tangent of the angle θ in Figure I.3 are

$$\sin\theta = \frac{a}{c} \qquad \cos\theta = \frac{b}{c} \qquad \tan\theta = \frac{a}{b}$$

Numerical values of these ratios are found in trigonometric tables.

TEMPERATURE

The scale of temperature most familiar to residents of North America is that of Fahrenheit: The freezing point of water is 32°F and the boiling point, 212°F. Most countries that use metric units also use the centigrade (or Celsius) temperature scale on which water freezes at 0°C and boils at 100°C. The lowest temperature that can be produced by any means is called absolute zero. At a temperature of absolute zero all of the thermal energy has been removed from a body. In centigrade, absolute zero is at −273°. On the *absolute temperature* scale, zero is taken at −273°C. Water freezes at 273°K and boils at 373°K. The "°K" stands for degrees Kelvin.

Appendix II
Geologic Time

The earth is about 4500 million years old. The oldest rock now exposed at the surface is about 3500 million years old; study of the surface rocks reveals many events which have occurred since that time. The method of determining the age of rock by radiometric dating is explained in *The Chemistry of the Earth* by K. K. Turekian.

Before the radiometric method was available, the relative ages of rocks were deduced from study of the fossils they contained. From time to time there were marked changes in the character of the plants and animals which inhabit the earth. Each of these changes marks the end of a *period.* Successive periods have been named and arranged in order to create a scale of relative time. This scale is shown in Table II.1. Also shown are the corresponding ages, determined radiometrically. Note that the scale of time determined from fossils ends at about 570 million years before the present. Only radiometric dating can be used to fix the chronology of the older rocks.

Reference

EICHER, D. L., *Geologic Time.* Englewood Cliffs, N.J.: Prentice-Hall, 1968.

Table II.1 Geologic Time Scale

Relative Duration of Major Geologic Intervals	*Era*	*Period*	*Epoch*	*Duration in Millions of Years (Approx.)*	*Millions of Years Ago (Approx.)*
Cenozoic	Cenozoic	Quaternary	Recent	Approx. last 5,000 years	0
Mesozoic			Pleistocene	2.5	2.5
Paleozoic		Tertiary	Pliocene	4.5	7
			Miocene	19	26
			Oligocene	12	38
			Eocene	16	54
			Paleocene	11	65
	Mesozoic	Cretaceous		71	136
		Jurassic		54	190
		Triassic		35	225
	Paleozoic	Permian		55	280
		Carboniferous: Pennsylvanian		45	325
		Carboniferous: Mississippian		20	345
		Devonian		50	395
		Silurian		35	430
		Ordovician		70	500
		Cambrian		70	570
Precambrian	Precambrian			4,030	4,600

Formation of earth's crust about 4,600 million years ago

Millions of Years: 0, 50, 100, 150, 200, 250, 300, 350, 400, 450, 500, 550, 4,600

Index